AUDITING AI

The MIT Press Essential Knowledge Series

A complete list of books in this series can be found online at https://mitpress.mit.edu/books/series/mit-press-essential-knowledge-series.

AUDITING AI

THE MARQUAND HOUSE COLLECTIVE:

MARC AIDINOFF
LENA ARMSTRONG
ESHA BHANDARI
ELLERY ROBERTS BIDDLE
MOTAHHARE ESLAMI
KARRIE KARAHALIOS
J. NATHAN MATIAS
DANAÉ METAXA
ALONDRA NELSON
CHRISTIAN SANDVIG
KRISTEN VACCARO

The MIT Press | Cambridge, Massachusetts | London, England

The MIT Press
Massachusetts Institute of Technology
77 Massachusetts Avenue
Cambridge, MA 02139
mitpress.mit.edu

The MIT Press would like to thank the anonymous peer reviewers who provided comments on drafts of this book. The generous work of academic experts is essential for establishing the authority and quality of our publications. We acknowledge with gratitude the contributions of these otherwise uncredited readers.

This book was set in Chaparral Pro by New Best-set Typesetters Ltd. Printed and bound in the United States of America.

Library of Congress Cataloging-in-Publication Data is available.

ISBN: 978-0-262-05172-9

10 9 8 7 6 5 4 3 2 1

EU Authorised Representative: Easy Access System Europe, Mustamäe tee 50, 10621 Tallinn, Estonia | Email: gpsr.requests@easproject.com

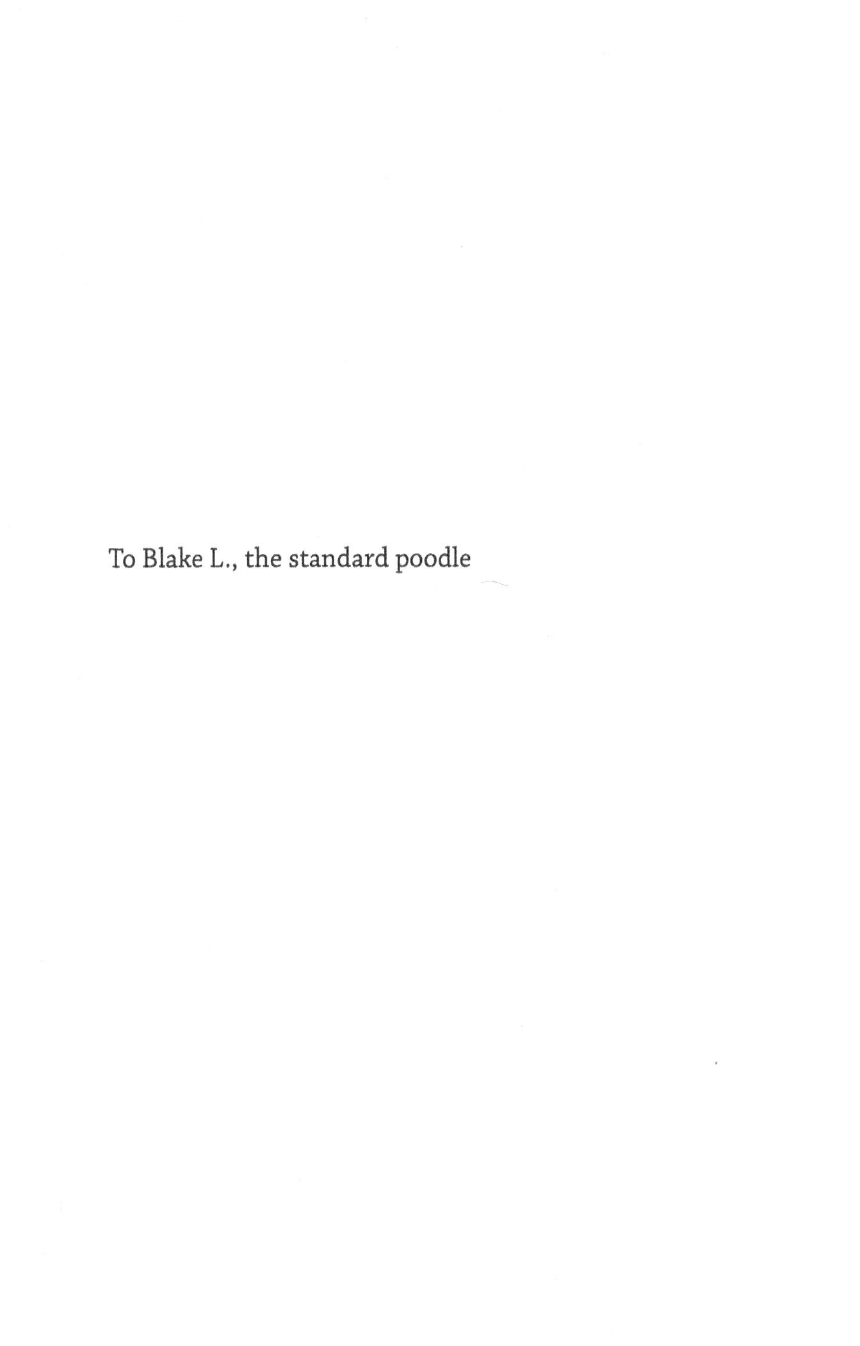

To Blake L., the standard poodle

CONTENTS

SERIES FOREWORD

The MIT Press Essential Knowledge series offers accessible, concise, beautifully produced pocket-size books on topics of current interest. Written by leading thinkers, the books in this series deliver expert overviews of subjects that range from the cultural and the historical to the scientific and the technical.

In today's era of instant information gratification, we have ready access to opinions, rationalizations, and superficial descriptions. Much harder to come by is the foundational knowledge that informs a principled understanding of the world. Essential Knowledge books fill that need. Synthesizing specialized subject matter for nonspecialists and engaging critical topics through fundamentals, each of these compact volumes offers readers a point of access to complex ideas.

1

INTRODUCTION

Transplant surgeon Peter Stock probably never expected to face a crowd of two hundred people arguing about whether or not a computer system could help save lives. When Stock finished his training in the 1980s, the idea of using computers to guide life-and-death decisions was the stuff of science fiction. But two decades later, when Stock found himself in a St. Louis conference room discussing organ allocation in the United States, the question facing him was not whether computers would make these decisions—it was how.

As chair of the US Kidney Transplantation Committee, Stock knew how much the possibility of a kidney transplant meant to every person with damaged kidneys. Kidneys filter toxins from the human bloodstream, and kidney failure can kill someone within weeks or even days. Roughly one in seven Americans, 37 million people, have some form of chronic kidney disease.[1] More than 100,000 people are on the kidney transplant waiting list each year, but only some 21,000 kidneys are available for transplant.[2] The waiting list gets longer each year.[3]

In 1984, the US Congress established the Organ Procurement and Transplantation Network to find a way to direct donated kidneys to people who need them most. At first, doctors sent kidneys to the nearest person in need. Then, as kidney transplants became safer and more routine, groups of patients, doctors, and politicians called for a fairer system. By the early 2000s, statisticians had developed an automated transplant prioritization system called LYFT that predicted how much longer a person could live if they received a donor kidney (LYFT an acronym for "life years from transplant"). But not everyone agreed with this system. Elderly patients alleged that LYFT put them at a disadvantage by prioritizing younger patients.[4] Communities of color and those facing poverty argued that it prioritized wealthier, more privileged patients. Everyone wanted to know how LYFT actually worked.[5]

As chair of the Organ Procurement and Transplantation Network's Kidney Transplantation Committee, Stock knew he needed to come to the debate equipped with both data and compassion. Stock's committee opened the meeting in St. Louis by presenting a series of evaluations of the LYFT system. The committee hoped these independent system analyses could quell people's concerns and inform a search for something better.[6] The evaluations Stock's group presented—audits of the LYFT system—sparked a series of analyses that identified disparities across patient populations, measuring differences based on age, gender, geography and race. This work informed decisions that would help bring greater equity to the use of this complex system for years to come.

Decision-making systems like LYFT represent a key component of what we now call artificial intelligence (AI). From the algorithms that direct lifesaving treatment to the software that filters our emails for spam, AI systems are now powering—and changing—countless aspects of our lives. Wherever AI systems are being used, people ask the same questions that Stock's committee did in 2009: Does it work? Is it fair? How will it affect my well-being, my safety, my livelihood, and my community?

To have informed debates about AI, society needs reliable information to help infer what goes on behind the screen—evidence that experts gather through a process called an *AI audit*. Such audits have now become common and sometimes even legally required in fields ranging from employment law to social media. Experienced AI auditors, like the authors of this book, are excited by the potential for audits to help leaders and the public know when to use (or not use) AI systems, improve public trust in AI, and guide improvements over time.

For most people, AI audits are new territory. Maybe an AI company has approached you with a system it is selling, or someone has complained about a product your own company has deployed. You might work at a nonprofit, be a journalist, or be a concerned member of the public who has heard about a proposed AI system and wants to understand and assess it. You may be an expert in your field, but AI may be new to you; you may never have commissioned or had to interpret an AI audit. This book is for you.

Rigged Reservations

While AI has captured the public imagination in the twenty-first century, the story of the first audits of computers started half a century earlier, amid debates over airline reservations.

Before the 1970s, reserving an airline ticket required hours of phone calls to book every passenger's seat. To buy a ticket, passengers would visit a travel agency, where an agent had to call an airline booking center to learn if a seat was available. Travel agents regularly spent hours on the phone booking travelers' tickets. At the other end of the line, airline staff would reserve seats by inserting push pins into a "seat board" and writing passenger information in a paper ledger.

Computers changed everything about airline booking. In 1960, the release of the groundbreaking SABRE system

(Semi-Automated Business Research Environment) computerized airline reservations in the United States for the first time (figure 1). Now travel agents could use a SABRE computer terminal to check for flights with available seats and book a plane ticket in just a few minutes.

But along with these gains in efficiency, new problems emerged. Some airlines noticed the SABRE system was behaving strangely. Republic Airlines alleged that its flights mysteriously disappeared from search results. Braniff International Airways alleged that passengers who booked with SABRE were then unbooked and switched to American Airlines flights. Continental Airlines claimed to have lost up to $55 million a day in business because its flights were absent from search results. The first booking suggestions on SABRE's screen were frequently on American Airlines flights—the same airline that owned and operated the SABRE reservation system for the entire industry.[7]

Republic, Braniff, and other American Airlines competitors did not know how the SABRE system worked, so they conducted an early AI audit. They compiled evidence of its biased behavior and pushed the US Civil Aeronautics Board to launch an investigation in 1983.[8] But rather than defending the system as fair, when then-president of American Airlines Robert L. Crandall testified before the US Congress, he instead declared that SABRE favored American Airlines flights intentionally! Crandall said, "The preferential display of our flights, and the corresponding

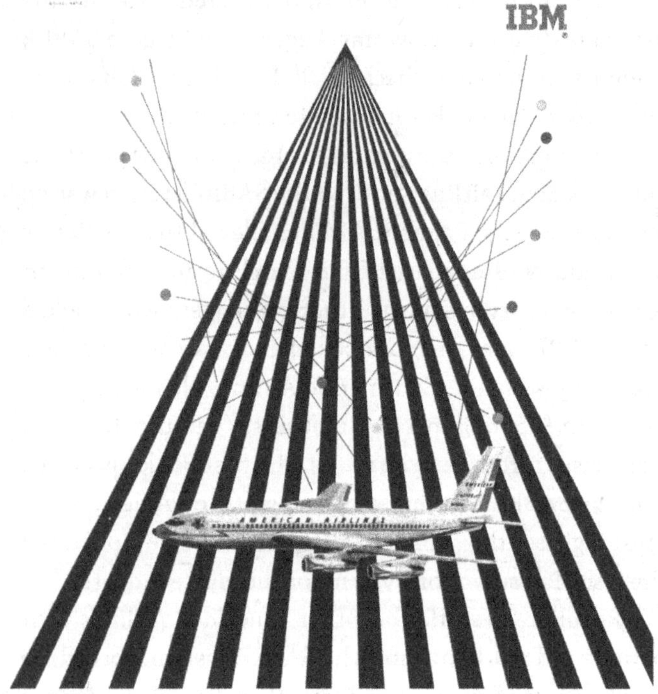

Figure 1 This 1961 advertisement highlights the partnership between IBM and American Airlines to produce a computerized airline reservation system, a milestone in the history of computing. American's competitors found the order of its search results suspicious, claiming IBM's system unfairly favored American's flights.

increase in our market share, is the competitive raison d'etre for having created the system in the first place."[9] He wondered why anyone would expect American Airlines not to build a system that favored itself.

Later authors identified this as a more general perspective on computers and automation and named it after the aggrieved CEO, christening it "Crandall's Complaint." Stated in more general terms it is the objection: Why bother to build an expensive AI system if you can't bias it in your favor? The implication of this perspective is: Without regulation, wouldn't we expect all systems to be biased in someone's favor?[10]

Crandall's justification did not go over well. In response, the Civil Aeronautics Board banned systems like SABRE from preferentially treating specific airline carriers and even gave the problem a name: *display bias*. The board ruled that the flight display algorithms—the set of steps performed by the SABRE system—must be made public. Any interested person could mail the company a written request and receive the exact criteria used to order SABRE's flight search results. This marked the first time in history that a regulation explicitly referenced an algorithm.

A version of SABRE is still in use today, providing the booking system for over four hundred airlines. As we will see in this book, contemporary life is heavily influenced by digital technologies like SABRE that help us find information, apply for jobs, rent apartments, choose videos to

The CEO wondered why anyone would expect American Airlines not to build a system that favored itself.

watch, select skincare products, and so on. Increasingly, AI systems are also playing a role in public agencies; in social services, healthcare, housing, and the courts, AI systems are making decisions that can change people's lives for better or for worse.

Unlike in the 1980s, however, people today cannot mail a request to the company and receive an explanation of the deciding details of these systems. The Civil Aeronautics Board no longer exists, and its display bias regulations have been repealed.[11] And even if there were such regulations today in other industries (there are not), the development of more capable AI means that the crucial internal logic powering our search engines, job sites, social media platforms, chatbots, and social services would not fit in an envelope.

What can be done to identify, prevent, and respond to misbehaving AI systems? This book will introduce and explain AI auditing, a valuable approach to answering such questions.

Targeted Discrimination

In 2022, several decades after SABRE, the US Department of Justice (DOJ) found itself dealing with another case of alleged computer bias. The case involved the social media giant Meta (formerly Facebook), whose site at that time served nearly two billion users.[12] Most people are familiar with Facebook as users of its social media platform, but its parent company Meta also operates an AI system that enables advertisers to place ads uniquely targeted to specific kinds of people that it thinks might be interested in those ads. Would-be advertisers can log in to a web page and purchase ads, uploading an ad image and selecting the type of audience they want it delivered to using categories like "animal lover" or "interested in beauty pageants."

A few years earlier, in 2016, journalists at the nonprofit media outlet ProPublica had published a bombshell investigation: Facebook's interface allowed advertisers to target—and exclude—audiences of different "ethnic affinities," such as "African American," "Asian American," and "Hispanic" (figure 2).[13] This was true even for ads related to housing or employment, a startling revelation since US federal law explicitly prohibits racial discrimination in housing and employment advertisements. The ProPublica investigation kicked off a multiyear saga including litigation brought by Facebook users against the company, investigations by multiple state governments, and eventually a lawsuit filed by the DOJ.

Throughout the nearly seven years between the initial story and when the DOJ secured a settlement in 2022, Facebook repeatedly insisted the issue had been addressed, or that the allegations were flawed. But every time journalists and academic researchers audited the ad system, they

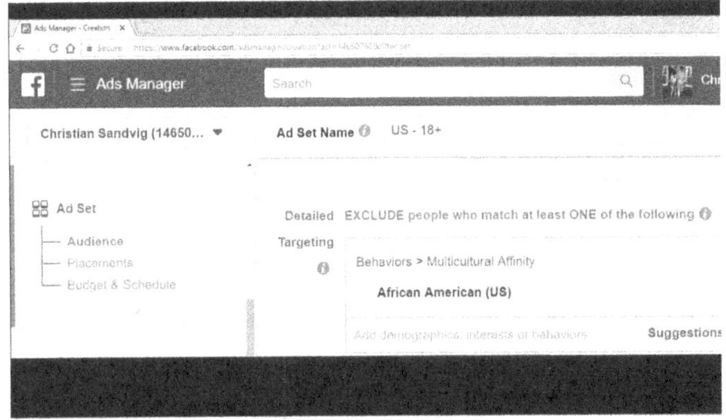

Figure 2 The Facebook Ads Manager interface once included a feature allowing advertisers to exclude people from seeing housing and employment ads based on "multicultural affinity."

found new evidence demonstrating that the platform enabled discrimination in areas like housing, credit, and employment which are covered by civil rights laws.

The case finally ended with a ruling that Facebook would pay the maximum penalty under the Fair Housing Act (or Title VIII of the Civil Rights Act of 1968)—a paltry sum of just over $115,000.[14] But the settlement also required Facebook to remove the functionality that led to the issue and to undergo routine, independent audits to proactively ensure that illegal ad targeting was not occurring on the platform.

AI auditing is a powerful method for understanding AI systems wherever they affect people's lives. Since SABRE,

AI audits have helped competitors ensure a fair marketplace, informed important public dialogue about lifesaving treatments, and alerted democratic institutions to AI systems that have crossed legal or ethical lines.

* * *

In all three cases discussed in this chapter, leaders and the public sought clarity about an opaque system and conducted research to study the problem. Sometimes the investigation was started by doctors and patient groups (as with kidney allocation), sometimes by competitors and the government (as in the SABRE case), and sometimes by journalists (as in the Facebook ads example). While the SABRE incident involves a firm deliberately placing a thumb on the scale of its computer systems, it is not uncommon for algorithm developers to be unaware of difficulties or the correct solution. In every case, audits revealed the real-world impacts of the AI system and prompted the search for improvements.

2

WHAT IS AI AUDITING?

Before discussing AI audits any further, we must first define what we mean by "AI systems." Algorithms—computational processes formalized in code, often but not always driven by data—are ubiquitous today at the operational core of AI systems. All automated systems in which algorithms are integrated can be termed *AI systems*. Although there are narrower definitions, we use this one because the accepted use of the term "AI" has grown impossibly broad. We use such systems in consumer settings (booking flights, targeting advertisements), educational ones (teaching and assessing students), and even in making life-or-death decisions such as prioritizing patients receiving medical care or helping judges make criminal sentencing decisions.

Why AI? If it seems to you that every digital service these days brands itself as AI, you are not alone. Industry leaders promise that, through automation, AI will improve efficiency and productivity, enhance decision-making, and augment human capabilities. Over the last decade, technological developments have taken AI from a niche academic field to a widely applicable foundation that can be deployed for seemingly endless uses. For all their purported promise, however, many AI systems have significant—and often silent—flaws. Whether due to unintended biases or outright malicious intent, they are far from infallible.

We define an AI audit as a *rigorous and systematic evaluation of the core claims, expectations, or standards pertaining to an AI system.*[1] Those standards may be based on what the law requires. The Fair Housing Act, for instance, was the law that Facebook violated when it allowed advertisers to exclude people, based on race, from seeing apartment rental ads. In other cases, we expect a system's standards to adhere to widely held social norms. Social media companies are notorious for inferring information about us based on our browsing habits, a practice that, for many users, violates norms and expectations surrounding privacy, even though it is not illegal. The deft auditor begins with the mindset of a skeptic, looking for evidence to prove that the system in question does not meet a given standard, including for accuracy, compliance with laws, and identification of potential risks.

The above definition is intended to capture the gravitational center of AI audits, describing the common thread

The deft auditor begins with the mindset of a skeptic.

found in *most* (but not all) strong AI auditing work. In the next chapter, we will explore the steps of a classic AI audit in more detail and discuss variations on that form.

Auditing: From Taxes and Landlords to AI

Readers will likely be most familiar with use of the word "audit" in financial settings. In the United States, individuals and small businesses can hire accountants to audit their finances. The Internal Revenue Service (IRS)—a government agency—also carries out audits. The latter is governed by tax code and has a more formal structure, from clear goals to potential financial penalties.

Audits (minus the "AI") are rigorous and systematic evaluations of an entity's or actor's claims, expectations, or standards. They are used across the private sector to assess whether a company adheres to laws and regulations governing its industry. US government entities, from the Food and Drug Administration to the Environmental Protection Agency to the Securities and Exchange Commission, encourage and sometimes require companies to undergo third-party audits to assess the effects of their business practices on public health, consumer interests, and the environment.

But other kinds of audits can affect the public interest and how private companies do business. In the early twentieth century, auditing emerged as a response to the problem of "redlining," specifically the practice of landlords and lenders who excluded Black people from certain housing and neighborhoods. Communities affected by redlining decided they needed a way to convince others that this was a pervasive problem. They needed evidence that was compelling and easy to understand, so local groups of tenants and homebuyers concerned about racial discrimination in housing banded together to establish themselves as fair housing committees.

The committees would send two people, a Black person and a White person, to ask a real estate agent or a landlord about apartments and houses on the market. They would do this test at about the same time, with both people recording how they were treated and what properties they were shown. If the landlord invited the White tester to tour an apartment but told the Black tester that no units were available, the committee would take note.[2] Perhaps inspired by tax audits, they called each series of tests an *audit study*.

After 1960s Civil Rights–era laws made housing discrimination illegal, these audit studies became a way to enforce antisegregation laws, as these audits were "designed to be so simple that even a lawyer or judge can understand them."[3] The basic structure of these tests was then adapted to other situations, like the job market. In another series of studies, a man and a woman tester with the same qualifications would apply for the same job at the same time.[4] These studies demonstrated pervasive discrimination in hiring (see figure 3).

The structure of these audit studies is also used in today's AI audits. Social science and activist audit studies systematically provide different inputs into a system (e.g., the tester's race) and examine the outputs (e.g., the landlord's reply) to understand whether the system is performing well against a given standard (e.g., the Fair Housing Act's nondiscrimination rules).

After decades of these audits, investigators began sending documents, rather than trained agents, to apply for jobs or housing. In one memorable audit (titled, "Are Emily and Greg more employable than Lakisha and Jamal?"), auditors sent nearly 5,000 fictional resumes to over 1,300 job postings, eventually finding that resumes whose names suggested the applicant was White received 50 percent more interview callbacks than those whose names suggested the applicant was Black.[5] These audits were used for two different purposes. Sometimes, the government, nonprofits, or academics used large-scale audits to systematically collect evidence to draw statistical conclusions about a social problem (i.e., the prevalence of housing discrimination). Other audits were conceived by prosecutors and tenants' rights organizations to point the finger at specific landlords or hiring managers. In both

Continuation of
HELP WANTED, MEN
Other Help Wanted, Men
In Preceding Section

HELP, MEN 24

ENGINEERING

OVERSEAS

TECHNICIANS
SPECIALISTS

IN

MICROWAVE COMMUNICATIONS

FOR INFORMATION
CALL R. S. TOWNEND
296-4571
PROFESSIONAL
STAFFING SERVICES
919 18th st. nw Suite 620. Pers.

ENGINEER—For lge. downtown apt. bldg. to assist. Chief. Call Chief Engineer, 293-1500.

ENGINEER—To work in power plant. 1st class lic. required. Apply Wash. terminal co., rm. 356, Union Station. E.O.E.

HELP, MEN 24

EXECUTIVE TRAINEE—$600 mo. BIG FAT ZERO. Have you been scoring a big fat zero lately? Everything you touch seems to turn out wrong? This needn't be. Maybe you're on the wrong track. Perhaps you are doing something wrong and can't put your finger on what it is?? Perhaps we can help you as this company is seeking executive material to start out 'learning the ropes' from the bottom up—advance as you absorb and learn. Are you good at handling people? Can you be discreet? Are you draft-exempt? This company insists that you keep your personal affairs out of the office—that you do not socialize with your subordinates. You must possess the all-important trait of being realistic as well as plain ole common sense. If you are a well-disciplined person, would like to earn top dollar, DON'T DELAY, CALL DEE DAY, 296-1560, Premiere Personnel, 1725 Eye St. NW, Suite 304, Premiere Bldg.

EXECUTIVE
OPPORTUNITY

We're looking for a man of executive character who is capable of

Figure 3 This 1970 issue of the *Washington Post, Times-Herald* shows jobs advertised as being available only to men. Illegal gender discrimination in employment led activists to begin in-person audits of hiring procedures.

cases, though, the best audits drew conclusions about *patterns* of behavior, not singular or specific decisions.

Just as researchers interested in employment discrimination realized they could use the US Postal Service and written resumes to speed up audits (which they called *correspondence audits*, because they were conducted via communication by letter), computer scientists recognized that computer programs could stand in for human testers and printed resumes. They called these automated tests *algorithm audits*.[6]

The growth of AI led to a shift in terminology, and today, we refer to algorithm audits as AI audits (although the terms are used interchangeably). We use AI audits to systematically collect evidence about an AI system's patterns of behavior. We often do this without full access to the system's inner workings, but the audits are still useful. The fair housing committees of the mid-twentieth century had no access to the internal workings of a landlord's brain or business strategy, but they were still able to produce evidence of biased housing decisions.

AI audits often investigate the same problems as these historical audits, such as discrimination, but they have also been adapted to many other situations: The earliest audit in computing (at the SABRE airline reservation system) was looking for anticompetitive behavior, while today's audit topics also include vital issues of AI safety, routine quality control testing, and more.

The fair housing committees of the mid-twentieth century had no access to the internal workings of a landlord's brain or business strategy, but they were still able to produce evidence of biased housing decisions.

AI Audits Take Off

In 2016, the Pulitzer Prize–winning investigative journalist Julia Angwin and her team at ProPublica released what went on to become one of the most widely discussed AI audits in history. The journalists studied a system called COMPAS (Correctional Offender Management Profiling for Alternative Sanctions), which was designed to aid judges in making sentencing and parole decisions for people convicted of crimes.[7] COMPAS, developed by a company called Northpointe (now equivant Corrections), promised a fair way to help determine if a person convicted of a crime would be likely to be arrested again. Judges in jurisdictions using COMPAS would take its recidivism scores into consideration when making sentencing decisions that carried extraordinary consequences for defendants. Research had already shown that the status quo—a judge assessing a defendant and deciding if they should be allowed out of jail—was a deeply flawed approach. COMPAS promised to solve this with its data-driven AI system.

When Angwin and her team published their investigation, the headline read: "Machine Bias: There's software used across the country to predict future criminals. And it is biased against blacks."[8] Angwin's team found that COMPAS incorrectly predicted that Black defendants would offend again at almost twice the rate of White defendants.

The system also mislabeled White defendants as "low-risk" more often than it did Black defendants.

Here was an ideal example of an AI audit in action: a methodical, well-organized evaluation that tested the accuracy, riskiness, and compliance of a system based on a stated standard. It was easily understandable by the general public and by experts, with a clear narrative and publicly available methods and data. It brought the idea of algorithmic discrimination more fully into the public consciousness and taught people to question the infallibility of a seemingly objective system. The publicity around the COMPAS audit demonstrated the importance of getting these systems right and resonated with ongoing civil rights advocacy focused on the rise of mass incarceration.

The importance of addressing unjust sentencing, and the COMPAS dataset Angwin's team made available, inspired many subsequent audits of the same situation. One study investigated the extent to which the scores affected judges' decision-making (since COMPAS scores were one of many pieces of information available to judges).[9] Others addressed fundamental questions of outcome: Did these scores, whether accurate or not, inadvertently support an inherently unjust sentencing and bail system?

Several COMPAS audits also underscored the pivotal role of data that is used to "train" AI. The inputs given initially to a system from which it can infer patterns are known as the *training data*. Training data are sometimes described as a set of examples from which AI develops rules or behavior. This phrase comes from machine learning and is meant to distinguish data used for other purposes such as validation or testing. Like any AI system, COMPAS was trained using data from the real world. In Northpointe's case, that training data was from real court decisions—where human bias was heavily present. It was little wonder, then, that COMPAS was biased too.

What Can AI Audits Do?

A well-executed AI audit asks a question and provides an answer. For instance, some teachers (including Karrie Karahalios, one author of this book), have audited AI systems used in university classrooms to give students automated feedback on their work. While many educators are excited about the potential of AI to augment their existing teaching by providing more feedback quickly and affordably, these systems also make mistakes. In this case, an AI audit conducted by Karahalios and her team answered questions such as, "What kinds of errors are most common?" and "When there is an error, how does it affect student learning?" The questions answered in a useful audit often go beyond the computers involved, considering real-world consequences and the people who are impacted by a system.[10] AI audits can include both technological and human dimensions and consider them together. Auditors can ask questions that combine what a machine does *and* what a human does in response.

Audits That Determine the Fate of an AI System

When they receive publicity, audits are sometimes framed as one-time, momentous investigations or scandals. Companies like PredPol (now Geolitica) and Palantir have offered law enforcement agencies expensive AI systems that promise to use data from past crime reports to predict when and where crimes will take place in the future. Criticism of these products has come from many quarters: Based on their general knowledge of crime and AI, criminologists and computer experts alike have found some claims made by these companies to be dubious. Some law enforcement authorities have said that knowing the approximate location of a possible future crime is not valuable information since police already know which areas have higher crime rates. Community activists have argued that the data being fed into these systems contain racial biases from decades of racism in policing practices and will therefore produce more of the same.

In this case, an AI audit allowed researchers to investigate concerns about so-called predictive policing systems and reformulate them as testable questions: How accurate is predictive policing? When it is accurate, does knowing the location of future crimes help police prevent them?

Audits of predictive policing systems revealing racial biases produced evidence leading several police departments across the United States to stop using predictive policing altogether.[11] In chapter 5, we discuss what happens when an audit prompts an agency or company to abandon its AI system. However, audits may not be the final judgment for a system; instead, they may be the start of a different kind of investigation.

Audits as Tripwires

The results of an AI audit are one among many tools for holding otherwise mysterious computer systems accountable. Sometimes, AI audits can ask simple questions in the service of later posing more complicated ones. In antitrust law, a widely used benchmark called the Herfindahl–Hirschman Index (HHI) provides a crude measure of industry competition. On its own, the HHI does not determine if antitrust laws have been broken. But a high HHI number can be used to justify additional scrutiny of a merger.

In a similar vein, AI audits can be used as tripwires to indicate that a problem may exist and that more scrutiny is needed, even when further investigation requires the use of other methods. Some AI audits are tripwires for legal proceedings, even serving as the basis for further investigations by government regulators. The audit identifying the potential for racial discrimination in Facebook ads described in the previous chapter is one example of an audit serving as a legal tripwire. That audit (along with subsequent follow-up audits) eventually triggered an investigation by the US Department of Housing and Urban Development and was presented as evidence in a legal complaint alleging that Facebook had enabled racial discrimination by advertisers.[12] As happens in legal cases, once proceedings were initiated, government regulators were entitled to receive more information from Facebook than the external auditors had access to. In this way, the external audit was a critical first step enabling further investigation.

Audits as Continuous Oversight

Rather than a momentous scandal, audits can also be part of a regular routine to help ensure the accuracy and reliability of AI systems. Many examples we have provided so far involve auditing a system after it has been deployed. However, anyone building or adopting a new system should solicit an audit and conduct it repeatedly to investigate how the system's performance or conditions change over time. Just as food is tested and graded for quality, giving us Grade A eggs, AI auditors can play the role of farm inspector, routinely checking that an AI system has not degraded in quality or developed new problems. This is particularly critical for AI systems that are constantly evolving, like generative AI.

The Curious Case of Google Image Search

In 2015, researchers noticed that Google Image Search results exaggerated stereotypes about gender roles and occupations. Searches for pictures of professions where women are underrepresented, like "doctor" or "construction worker," turned up very few images featuring women. The researchers' audit confirmed that the search results showed even fewer women than were known to work in those occupations, according to data from the Bureau of Labor Statistics. Even worse, when images of women were shown, they were much more likely to be unrealistic caricatures, a phenomenon AI experts dubbed "the sexy construction worker problem."[13]

Google never formally addressed the findings, even after they were made public. Google's lack of response, and the fact that it is normal for search results for the same search term to change over time, left an opportunity for auditors to try again. Four years later, in 2019, the Pew Research Center replicated the previous audit. The findings continued to reveal an imbalanced portrayal of occupations, especially for women. This could discourage women and girls from pursuing a wide range of careers and exacerbate discrimination and harassment for those already employed in these fields.

The audits did not stop there. In 2021, one of the authors of this book, Danaé Metaxa, led another AI audit investigating gender as well as racial stereotypes in search results. Unfortunately, they found that stereotypes persisted.[14] While disappointing, the findings highlighted the importance of repeating these investigations rather than simply trusting a company to fix its system. In other words, audits can also ask: Is the problem improving or getting worse? AI audits can test whether intended system improvements have occurred or undertake routine monitoring of a system where frequent changes are expected. We discuss these dynamics in greater depth in chapters 3 and 5.

Boring and Valuable

The Google Image Search story above could have been a much less interesting story. All of the audits could have found that the search results offered totally representative or even balanced portrayals of women in male-dominated professions. But these would still be examples of valuable audits. An egg inspector's work is surely boring if all the eggs examined are Grade A, yet we are happy to have these boring inspections as they promote confidence in food safety and the food supply. There is value in audits that are boring or banal, in the sense that they find no problems. This is even more important in safety-critical or high-risk systems, where a regime of boring (or one might also call them "passing") AI audits that test and ensure the quality of the implemented AI should be routine.

What Can't AI Audits Do?

Although they are effective at answering many important questions, AI audits cannot solve every technical problem or answer every question about an AI system. Most AI audits are focused on what a system *does* instead of *how it works*.

As an aside, fully understanding how some systems work may be impossible. With any complex technology, it is common to find that both the people who built the system and those who use it struggle to completely explain how it works. Among AI experts, the word "explainability" refers to the ability to describe how a system operates in a way that is understandable to humans. Even when a system is explainable, many companies are often loath to reveal their true inner workings—a challenge of opaque industry behavior that even the most rigorous audits cannot wholly overcome.

AI Audits Often Cannot Open the "Black Box"

How does a particular AI system work? We may never fully know. AI systems are often described as "black boxes," meaning that the people who use them cannot see inside them to examine their internal operation. They just see the inputs (like search terms) go in, and the outputs (like search results) come out. Some systems are black boxes even to the people that own them or build them.

Imagine a young indie rock artist trying to break into the music industry. She might find that Spotify seems to feature numerous country musicians on its up-and-coming playlists, but rarely indie rockers. (This is not so far-fetched; in 2022, independent artists complained about a different bias, alleging that Spotify's music suggestions prioritized already-established artists over newcomers).[15] Commissioning an AI audit could confirm this trend—that Spotify's recommendations favor country music and disadvantage indie rock. The audit results might convince this artist that she should pivot to an entirely new genre to have a better shot at gaining publicity on Spotify. But behind the scenes, the reality might be quite different. Rather than prioritizing country music, Spotify might instead be accepting pay-to-feature artists on its so-called up-and-coming lists, with country musicians' publicity teams more likely to be paying Spotify to advertise them. In that case, even if she changes her sound, the indie rocker is still unlikely to have success.

But imagine another possibility: Sometimes, even the engineers who built an AI system don't know why it behaves as it does. Anthropologist Nick Seaver has shown that engineers who develop music recommendation systems are sometimes surprised and confused when confronted with the outputs of the system they work on, even though they built it.[16] Indeed, this surprise can also be a problem with non-AI computer software that is complicated and has been developed by many engineers. An AI audit can help draw inferences based on a system's observable behavior, even if none of us know what is going on inside. AI audits provide insights into these systems' behaviors, but not their precise causes. Audits are not the same as explanations.

AI Audits Do Not Guarantee Accountability

Revisiting our previous example about biases in Google Image Search, Google's lack of response or improvement to its tool despite a series of audits conducted over seven years highlights another limitation of AI auditing. Even when audits repeatedly identify important problems, they do not guarantee that an organization will address them.

Driving this point home, Google released a new AI system in 2023 called Gemini. The system used generative AI to produce text, images, and video in response to users' text prompts. Users soon noticed that Gemini had a diversity problem, but not the one you might expect. Rather than underrepresenting gender and racial minorities, it appeared to ensure their representation in its outputs, regardless of context. When users asked Gemini to produce, for instance, an image of German soldiers battling in World War II, it generated images of young Black men wearing military garb with Nazi insignia (figure 4).[17] These results were both historically inaccurate and glaring, given the White supremacist ideology of the Nazi regime.

Figure 4 In response to a request for a "1943 German soldier," this screenshot from Google's Gemini AI shows the historically inaccurate result of adjustments meant to improve gender and racial representation.

Clearly, Google overlooked the argument made in Metaxa's 2021 audit, in which the authors emphasized that "achieving greater representation in search may not always translate into desirable real-world outcomes." They wrote that anyone building tools like Google Image Search must "anticipate, measure, and evaluate the socially situated impacts of their work, lest such second-order effects of well-intentioned efforts do more harm than good."[18] AI audits identify problems, but they do not necessarily specify the right solutions or hold a system accountable.

Who Should Conduct AI Audits?

The goal of this book is not to dictate who can do an AI audit, but rather to help define what makes for an effective and credible audit. A cornerstone of a robust audit is that it is conducted by people or teams empowered to surface and articulate issues even when doing so presents problems for the AI system's owner. This type of independence is necessary to preserve the potential for AI audits to lead to genuine accountability.

Operations teams across industries and governments need to hire and support reliable auditors who are technically able and organizationally empowered to evaluate the core claims, expectations, or standards of AI systems. Currently, AI auditing builds on the expertise and authority of a wide range of entities. Government agencies, law firms, consulting firms, corporate auditors, civil society organizations, journalism outlets, industry researchers, and academics all have a hand in this kind of work. Some organizations have increased their staffing to develop distinct expertise on how to probe technical systems. Established consulting firms have also begun offering their services as auditors, and various websites have begun advertising recently invented auditing courses and credentials.

But there is a danger in a world where AI auditing is so professionalized that we have unwittingly incentivized a new bureaucratic class. We have no desire to create a new legion of mid-level technocrats armed with checklists and rubber stamps. Instead, we aim to support a healthy and diverse ecosystem of skilled AI auditors who recognize the real-life impacts of the technologies they are tasked with auditing.

Internal Auditors and Conflicts of Interest

Can people employed by or affiliated with a particular company or government agency conduct an effective audit? There is an ongoing discussion within AI auditing on what role—if any—internal auditors should play. As we lay out in chapter 3, a strong AI audit involves rigorous evaluation of a developer's or organization's claims about an AI system, usually by third parties independent of the system in question. Audits conducted by independent third parties

We have no desire to create a new legion of mid-level technocrats armed with checklists and rubber stamps.

are considered the gold standard in AI auditing because they do not have conflicts of interest that could negatively impact the audit.[19]

Companies routinely undertake internal testing to measure the efficacy of their products, but this is different from conducting an audit. When audits are conducted by the same entity that built and stands to profit from the AI system in question, faulty or problematic results can follow. The COMPAS system we described earlier in this chapter is one such example. Northpointe (creator of the COMPAS tool) had conducted its own tests ostensibly proving the system was unbiased. But ProPublica's independent journalists found otherwise in their landmark investigation.[20]

Even the mere suggestion of a conflict of interest can damage the credibility of an audit. In one example, a well-respected academic researcher published an audit of an AI-based job candidate screening tool built by Pymetrics. He conducted the audit with the permission and collaboration of Pymetrics.[21] Although the audit presented in the paper was described as independent, company employees were listed as coauthors of the publication. This raised red flags, triggering concerns about corporate capture of academic research.[22]

There are also many examples of internal audits that have been well received. These include a paper from internal researchers at Twitter (now X), following allegations about racial bias in an AI system that automatically cropped images featured in tweets. The tool repeatedly cropped images to focus on White faces, regularly leaving out people with darker skin tones. The Twitter researchers published the results of their own formal audit, confirming the bias and abandoning the system after their attempts to improve it were unsatisfactory.[23] Another example comes from Amazon, when investigative journalists, aided by anonymous sources, revealed that Amazon had audited its own resume scanning tool, found it to be biased against women, and discontinued its use.[24]

Some companies may choose to commission an audit. In this case, they bring in an outside firm to conduct an audit but work with that firm to determine the scope of the audit. Self-commissioned audits are important for companies to inform their own policies and for public transparency, but they accomplish different goals than independent third-party audits. Because independent audits typically disclose all findings—including negative findings that produce legal liability—they provide more transparency to the public and to regulators.

Company-led audits do have some unique benefits. Unlike auditors working independently, auditors working hand-in-hand with a company are more likely to have access to a complete set of data and information about the system's operation. Companies can also affirmatively collect the internal data necessary for particular company-led audits, where outside auditors may not have any access to key data. In some cases, this can lead to strong mechanisms for accountability through auditing.

In 2016, Airbnb commissioned a civil rights audit of its own platform in response to claims that it enabled discrimination against users based on race. The results were published, and ongoing updates were provided to the public as part of the company's efforts to increase transparency and presumably to burnish its reputation for being responsible and accountable. That initial audit led the company to launch Project Lighthouse, an initiative with the civil rights nonprofit Color of Change and others, to collect data to enable ongoing assessment of racial discrimination on the platform. Such a strategy can carry risks if an audit is perceived as audit-washing a system's problems, or when not enough detail is provided to enable outsiders to assess the audit's soundness. But if they are undertaken with care, using an audit as a launchpad to begin a public program of collaboration to address a particular harm is considered a best practice.

* * *

In this chapter, we defined AI auditing, describing how it was inspired by the fair housing committees of the 1940s and the subsequent audit studies during the civil rights movement. We showed how computers have adapted and changed the original tests of landlords and employers, noting the kinds of situations where audits are useful and perform well, and the tasks where audits are not the best choice. We also reviewed perhaps the most-written-about AI audit to date, that of the COMPAS system for predicting recidivism by people convicted of crimes.

3

THE STEPS OF AN AI AUDIT

As part of a graduate course at MIT, Joy Buolamwini and her classmates were building an interactive art installation called UpBeat Walls. Their idea was a play on face painting—their computer system would allow people to control the colorful shapes and patterns projected on a wall by moving their heads. Excited to begin testing the tool after spending long hours programming, Buolamwini immediately ran into a snag. The off-the-shelf face detection technology her team relied on to detect faces did not work. She stood in front of the camera and tilted her head back and forth, but the system did not respond. Buolamwini, who is Black, instantly knew what was wrong.

A few years prior, while working on another project using face detection software, she had run into the same problem. During a Halloween weekend, while debugging the issue, she happened to try putting on a white costume mask she had lying around from the holiday.[1] It worked. Presenting the project to her friends confirmed her suspicions. The face detection software only worked on light-skinned faces.

Back in the classroom, Buolamwini was frustrated. On reflection, she could think of several experiences over the years when the face detection systems she used in her classes did not work on her face. How pervasive was this issue? And why had no one fixed it yet? Buolamwini wanted answers. She decided to conduct an audit alongside Timnit Gebru, a researcher at Microsoft Research. They named their audit "Gender Shades" and it became a canonical example in the field of AI auditing.[2] Confirming their suspicions, the audit found that face analysis systems were less accurate for dark-skinned female faces than light-skinned male ones.[3] In the next sections, we will use the Gender Shades audit and other audits as examples to introduce and walk through the procedure for AI audits in general.

The Five Steps of an AI Audit

All audits—from an investigation into facial recognition AI to an analysis of automated decision-making systems to recommend prison sentences—involve five basic steps: (1) identify the system to audit, (2) plan a systematic investigation, (3) collect data, (4) analyze the data, and (5) communicate the audit's conclusions (figure 5).[4] Decisions made at each step of this process influence the organizational challenges of doing the audit and the impact of the results.

Before You Begin, Clarify the Stakes

The Gender Shades audit became famous because of its clear design, surprising findings, and effective presentation: Buolamwini made a short video accompanying the audit demonstrating that a computer could not accurately label a Black woman's face—her own face. Gender Shades touched a nerve—automated analyses of a person's face are technologies that can change the course of a person's life.

Journalists have reported on wrongful arrests of people in New Jersey, Michigan, and Louisiana—all of whom were Black—on the basis of a bad match by a facial recognition algorithm. In 2019, Detroit resident Robert Williams was wrongfully arrested after police used an AI facial recognition system that misidentified Williams as the suspect in a luxury goods store robbery.[5] In the case of Williams' arrest, the prosecutor's office dropped the charges and prosecutor Kym L. Worthy cited the Gender Shades audit in the decision: "In the summer of 2019, the Detroit Police Department asked me personally to adopt their Facial Recognition Policy. I declined and cited studies regarding the unreliability of the software, especially as it relates to people of color."[6] The prosecutor noted that the

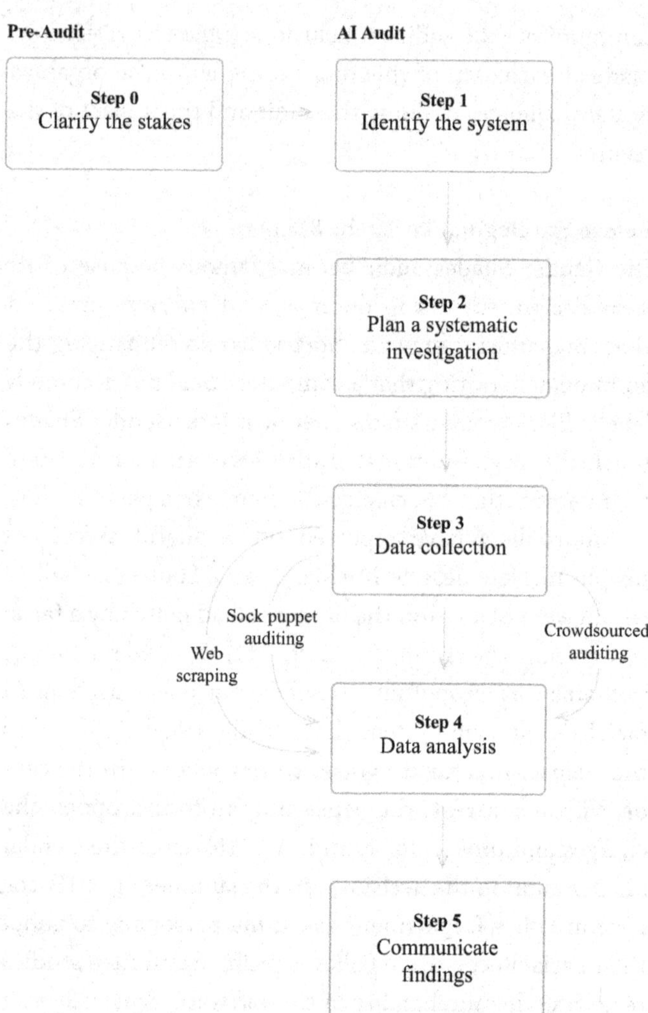

Figure 5 The five steps of an AI audit. Illustration credit: Shannon Yeung.

dismissal of the case did "not in any way make up for the hours that Mr. Williams spent in jail."[7]

Selecting an audit target with enormous real-world consequences was a major factor contributing to the downstream impacts of Buolamwini and Gebru's audit. Some of the companies that produced this technology admitted that they conducted no tests of their own similar to the Gender Shades audit before deployment.

In contrast, when Twitter received complaints from users about racial bias in the company's image cropping AI, the company investigated. Twitter's internal team of AI ethics researchers found that the AI system was cutting certain people out of photos—and it seemed to be biased against people with darker skin.[8] To guide auditors, they published a list of nine harms that were especially concerning to the company, from denigration and stereotyping to reputational and economic harms to the company's users. By clarifying the stakes, the team at Twitter provided a North Star that future audits of Twitter would follow.

Step 1: Identify the System

Identifying the AI system to audit can be harder than it might seem. What we call "AI" is usually a bundle of many components with interconnected parts whose boundaries are hard to pin down.

Consider the journey of an online ad from the advertiser that created it through to the ad appearing on someone's screen on Facebook. Along the way, the ad gets evaluated by different AI systems that compute how much to charge the advertiser, which interest groups it might be able to reach, and whether it follows the company's terms of service. If the ad passes these checks, it goes into an internal marketplace AI system (built from multiple AI systems) that weighs all the ads in circulation and decides which people see which ads.[9]

When auditors clearly understand the stakes, it is easier, in theory, to choose which parts of a system to investigate. But those choices can be more difficult when managers or outside auditors do not know where (and which) AI systems are being used. In those situations, a good place to start is to create a systems diagram.

For example, an equity research analyst named Dan Salmon created a diagram to understand Facebook's advertising market.[10] Tasked with advising investors on how much to spend on Facebook ads, Salmon realized that few companies actually understood how Facebook's systems worked. So he studied the structure of Facebook's ad market by reviewing public information about it and made a diagram of its component parts.[11] The diagram made it clear that anyone trying to understand Facebook's ad market also needed to understand the role of third parties called agency trading desks (ATDs) and demand-side platforms (DSPs), as shown in figure 6. Researchers who

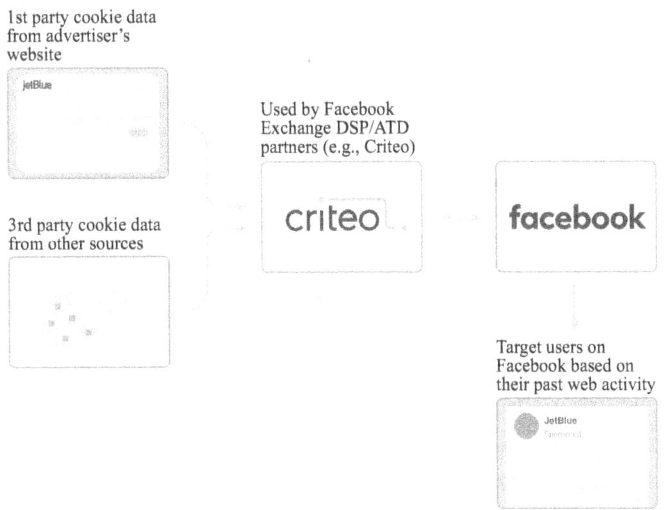

Figure 6 An example of a system diagram showing how Facebook's Ad Exchange worked. Diagrams like this help auditors determine which part of an AI system to target. This illustration is a much-simplified version of Dan Salmon's influential diagram. Illustration credit: Shannon Yeung.

subsequently audited Facebook's advertising algorithms were able to consider this information when deciding which systems to study. It seems only natural that if you want to look for problems in the process that decides which ads appear on Facebook you should study Facebook. In reality, Salmon's diagram shows that a number of other firms are involved when an ad appears and these companies might be the more appropriate target for an audit.

Step 2: Planning a Systematic Investigation

Like Buolamwini, everyone has had an experience using a computer when software did something they did not expect or found distasteful. As part of a research project on online advertising, Christian Sandvig, one author of this book, received ads asking, "So You're Going to Prison?" and "Is Your Spouse Considering a Divorce?" It's tempting to think the algorithm might know something about you that you don't know.

Why did the AI system responsible for targeted advertising show these ads? Although they are not sound evidence in and of themselves, specific individual experiences and personal explorations can serve as inspiration for audits. But to turn an anecdote or scattered set of experiences into an audit, one must be systematic.

What does it look like to be systematic? In one of the earliest AI audits, Harvard professor Latanya Sweeney was on Google, searching for a paper she had written to show to a colleague. She googled her own name. But instead of finding the paper, the first search result read: "Latanya Sweeney, Arrested?"[12] She was shocked. Sweeney had never been arrested, so why did the top ad on Google suggest she had been? As a Black woman, the experience left her suspicious that something systemic was wrong with Google's ad system.

To turn her individual experience into a rigorous audit, Sweeney used statistical data about baby names from the

But to turn an anecdote into an audit, one must be systematic.

National Bureau of Economic Research, birth records, and other sources to develop a list of over two thousand full names that are statistically extremely likely to be given to either a Black or White person.[13] Think of names like "Latanya" or "DeShawn," compared to "Kristen" or "Geoffrey." Querying Google with those names (the next auditing step we discuss below), she was able to turn her intuition into evidence. Black names were 25 percent more likely to turn up an ad suggesting an arrest record, compared with White names—regardless of whether anyone with that name had been arrested.

Step 3: Collecting Data

Even if you are not an auditor, you should still care about how data is collected and used during audits because these details have profound implications for the ethics, privacy, and legal decisions that make audits possible. The data collected must be appropriate for the intended audit questions and answers. Auditors must ensure the data captures the audit goal and is valid for the population it represents. If the collected data is unreliable or does not represent the appropriate audience, the audit results will not be deemed credible.

To answer her question about whether Black or White names received different Google ads, Sweeney, in the example above, didn't select names based on stereotypes; she used birth records and statistics. She then manually ran a Google search on each name in the dataset and documented the results that Google returned. While this approach systematically answered Sweeney's question, one might want to conduct a similar audit at a larger scale. Auditors have developed several methods to allow us to do exactly that.

Scaling Data Collection

Although Sweeney collected her data manually, she could have leveraged technology to test outputs for many more racial groups and nationalities. Many auditors today use computer scripts to automate and scale data collection.

One common approach to collecting data at scale is *web scraping*, a technique for downloading the contents of a web page. Web scrapers often pretend to be real internet users and operate without a target's permission. This method is as powerful as it has been contentious; in fact, some of the authors of this book have been involved in lawsuits to protect the rights of auditors to scrape data, as we describe in more detail in chapter 6. In other cases, software platforms that are audit targets provide programming tools designed to allow scripts or other software to interact with them: These are called *application programming interfaces* (APIs). Unlike scraping, when a company provides an API, it acts like an invitation: It allows outsiders to use scripts and other automation with the company's data. Anyone may be able to use an API, or the system's operator may require permission in advance. Scraping remains relevant because many AI systems do not offer APIs, and all APIs place limits on how they are used.

Some tasks require more information than can be collected with web scraping or APIs alone. Recall the classic housing audit in chapter 1, where a Black person and a White person were sent to apply for housing and then recorded how the housing agent responded. To run hundreds or thousands of tests quickly—many more apartment "visits" than the few a person could see in a day—auditors can create fake, automated profiles, called "sock puppets."[14] *Sock puppet audits* recreate the structure of traditional audit studies, but instead of sending real people, we send automated programs (sock puppets) to represent people of varying racial groups, nationalities, and genders, among other traits. Sock puppet audits are more complex than scraping because they usually involve a fake profile with a login for each imaginary tester. With systematic planning, automated data collection is faster, cheaper, and more consistent than the offline collection that inspired traditional audits, although sock puppets can fail to capture ancillary information that a human tester would likely observe and document.

In cases where real user data is paramount, another way of collecting data en masse is to ask regular people—and lots of them—to pitch in. In what are known as *crowdsourced audits*, auditors build tools that make it easy for people to donate their own data to auditing efforts. For instance, following reports that the travel booking website Orbitz was steering Mac users to higher-priced websites compared to Windows users, auditors built a browser plug-in that anyone could use to capture key data points about prices on booking websites.[15] Hundreds of real-world users installed the plug-in, which captured and sent that data to auditors. This allowed them to compare the prices shown to different people and identify other websites likely to be engaging in similar price steering and discrimination.[16] These audits have also been called *collaborative audits* because the users of a system are in a way also acting as the outside auditors or at least they can be thought of as being a part of the auditing team. Such efforts underscore the power of strength in numbers—by relying on the crowd, auditors can collect real data on a large scale.

Ethical Access

The different approaches to collecting data reviewed above pose different ethical and legal questions. While there may be ethical and legal concerns about how publicly available data about people are used, there are likely to be even more restrictions on data that are not public. Audits are traditionally carried out without the advance consent of the people involved: The potentially criminal, racist landlord in chapter 2 was not asked beforehand if he could please be investigated for violating the Fair Housing Act. This is an unusual ethical situation, and we discuss it in more detail later. Even without the legal requirement of consent, auditors have other ethical obligations, such as a duty to reasonably secure personal data they obtain in order to prevent future hacking or identity theft.

Because AI companies collect and make use of lots of personal data, that data often becomes part and parcel of an audit. As necessary data privacy laws proliferate at the state level in the United States, there may increasingly be legal limitations on audits collecting, using, and disseminating data without a person's consent. As of 2024, eighteen states have consumer privacy laws that impose limits on personal data collection and use.[17] While these laws vary, some may impose limits on audits that rely on personal information collected by companies, particularly if the auditors work in concert with those companies.

Finally, collecting data for an audit in an automated way can be taxing to the AI system being studied. Most AI audits test systems in ways that the target software was designed to handle: For example, Google Search is designed to respond to search queries from the public. But if auditors query Google Search too many times and too quickly, they can impose a cost on the system outside its normal tolerance, causing it to slow down or even crash. Irresponsible auditing could thus harm the audit target or other users, who might be unable to use the system as normal. Using too many sock puppets (or too much simulated data of any kind) can itself also change the operation of the system being tested. The costs of an audit are discussed in more detail in chapter 6.

Step 4: Analysis

In theory, computer-based data analysis can happen at the click of a button. But good data analysis is an iterative process that often requires back-and-forth conversations between many people and the iterative use of statistics.

Most high-quality audits will highlight the trade-offs inherent in these complex systems and carefully lay out the specific harms the system is causing. When the US Kidney Transplantation Committee set out to audit the LYFT system, it ended up recomputing the analysis to test the impact on many different communities, including elderly people, younger people, women, people in different geographies, and people of color. The committee only arrived at a final decision once all these groups were satisfied that the analysis accounted for the impact of the LYFT system on their lives.

Meanwhile, some compliance-oriented auditing may instead try to streamline and templatize the audit process. New York City saw an influx of such audits when it passed a law requiring employers to commission independent audits of software products and platforms used for hiring. Upon the law passing, vendors of AI hiring tools worked with auditing firms to create standard templates for the audits required by the city. When employers asked their vendors about producing audits, the vendors were ready to direct their customers to a list of auditors who could work from these templates.[18] But a thorough audit answering whether a hiring tool is legally compliant in its actual use likely needs more than a generic template result.

Step 5: Communicating Findings

The final step is communicating the findings of the AI audit to groups that need to know. If the audit is mandated by regulation, there will likely be requirements about what to report and to whom. In some cases, those requirements may be minimal, only asking auditors to report whether the AI system passed or failed different parts of the audit. But groups conducting audits should aim to communicate more than those minimal requirements.

In general, sharing more information about an audit offers several benefits: News of a successful audit builds trust in the system, details about the design of the audit allow people to better assess its results, and describing the particulars of an audit can help reduce criticism and potential backlash about missing information. Sharing audit results can have reputational benefits whether or not the audit finds a problem. When no problems exist, the company can celebrate its success and establish stronger trust in its system. If issues are found, negative publicity may follow, but if the company resolves the issue, this is an opportunity to highlight the strength of its processes. Amazon and Twitter revealed results from internal AI audits that showed problems in their systems, and they received positive press attention as a result.[19] Computer companies have a long history of voluntarily revealing security vulnerabilities or bugs in their own systems and repairing them. This too can yield favorable press coverage and other reputational benefits.

The Gender Shades audit that we described at the start of the chapter drew remarkable press exposure and public attention, thanks both to the compelling questions the researchers posed and the rigor of their methods. Dozens of news articles about that audit and those inspired by it generated pressure from the public—and from the US Congress—that led to real change. By 2020, Microsoft had stopped selling its facial recognition technology to law enforcement agencies and IBM had vowed to stop building the technology altogether.[20]

Some audit results are shared in academic publishing venues, on companies' websites, or in public records. Independent auditors often share their results with the press, benefiting the broader public. When audits are conducted by outsiders, it is common practice to share the results of an audit with the target system for comment before releasing it publicly.

Sample Audit: Does LLM-Generated Writing Discriminate Against Women?

Having explained the five steps entailed in many AI audits, let us work through an example to illustrate the entire process from start to finish. Recently, the development of large language models (LLMs), which generate long-form text based on a user's written prompt, has drawn a lot of excitement and criticism. A group of auditors including Lena Armstrong and Danaé Metaxa, two authors of this book, conducted an AI audit to evaluate one such model, OpenAI's GPT-3.5, for gender and racial biases, since such issues are widely known to plague similar AI systems.[21]

The researchers needed to select a system and question (step 1). They chose to study one of the most widely used LLMs of the moment—GPT-3.5, which at that time was the model underlying OpenAI's ChatGPT. They were curious about how employers might be using these models to aid in their hiring processes, drawing on media reports indicating that people had begun using LLMs to evaluate resumes. They posed an initial question: Was GPT-3.5 biased against particular racial and gender identities? That is, would names strongly associated with certain identities produce the same kind of employment biases that appeared in classic audits involving real employers? More concretely, was the model treating resumes for Black people and women differently than those for White people and men?

Next, they needed to plan a systematic process to answer that question (step 2). Other studies had identified gender and racial biases in software tools designed to assign scores to resumes. In addition to understanding bias when the model scores resumes, these researchers decided to include another approach that would provide a new kind of insight into the system's biases. So they wrote a computer program that asked the language model to generate resumes for fictional job applicants (step 3), providing groups of names that had strong associations with race and gender (e.g., Monique, Kenji, or Rodrigo). The group then analyzed the outputs (step 4) to check for any differences between racial and gender groups. They used GPT-3.5 to generate dozens of resumes for each name, to be sure they could draw conclusions about its behavior in general, rather than just gathering a few anecdotes.

A thoughtful reader might ask: Is this a realistic task for a large language model? The answer is clearly no. Real job applicants obviously would not ask a model to generate them a fictional resume and use it to apply for jobs. But the goal of this study was to better understand the AI's behavior when given an open-ended employment task with implications for job recommendations, evaluations, and hiring. Another audit could perhaps have the model edit people's real resumes to see what biases it introduces. Such an audit would allow the model less leeway than writing resumes from scratch but would more closely parallel possible real interactions with the system. There is no right or wrong audit design, just different costs and benefits to each option.

Using this method, the team created 320 resumes, each for a hypothetical job applicant whose name suggested a particular race and gender. They then manually reviewed the resumes produced, annotating them for eight possible biases and ensuring the manual labeling was reliable by statistically verifying that several members of the team agreed on each annotation. These biases included the number of years of experience the resume reflected, the industries the hypothetical applicant had worked in, any gaps in his or her employment history, and others.

After the extensive manual labeling process, the auditors took a step back and looked at the results across all 320 resumes. In aggregate, they found that prompting with a woman's name led GPT-3.5 to produce resumes that, on average, listed fewer years of work experience compared to those generated with men's names. Gender bias: confirmed. The language model seemed to think that a working woman would be more junior or have less experience than a working man. The results did not find similar issues with race but did reveal a new bias. Resumes generated using Hispanic names were much more likely than other resumes to indicate that the applicant was a Spanish speaker and was not a US citizen (writing, for instance, that he or she was a green card holder), and resumes generated using Asian names were more likely to list an Asian language. In other words, the audit revealed racial biases relating to nationality.[22]

Some of these results clearly don't reflect reality: For example, only about 3 percent of the US population holds a green card, and the majority of Hispanic people in the United States are not immigrants. None of these results are completely unexpected: Social biases against women abound in the workplace, as do stereotypes about Hispanic and Asian people being immigrants or not speaking English well. This audit revealed that a new AI system, suddenly widely used and included in many high-stakes settings, was likely to reflect these biases too.

Revisiting the initial audit question, the audit team was able to answer a range of questions about specific types of bias in this LLM and in several cases (but not all) found cause for concern. As an academic research team, the auditors chose to share this work in a technical research article (step 5), as well as with the general public through general-audience summaries on social media and by speaking to the press.

The example above summarizes the steps of a typical AI audit: The auditors begin by *identifying a target system* (GPT-3.5) *and question* (Is there race and gender bias in resume writing?). Auditors then *plan a systematic investigation*, framing their concerns as questions that could be answered by designing varied inputs to the system (In this case, the model was prompted with names chosen specifically to evoke job seekers from specific racial groups and genders.). They *collect data* (collecting model-generated resumes for each name) and, to answer their questions, *conduct an analysis* to compare the outputs between groups (e.g., the LLM using markers in a resume to link race/ethnicity with immigration status). They *communicate findings*, sharing what they learned with the research community, the general public, and even the company responsible for the system tested.

The Audit Variations

For the sake of simplicity, the steps above describe what we might call classic or prototypical AI audits. As we saw, these audits involve providing different inputs to a system and then examining its outputs. Some auditors think of this as the minimum definition of what is required to have an AI audit: An audit varies the inputs and checks the outputs. But there are variations on this classic structure that are widely and effectively used by auditors. We describe several of these below, to give you a sense of the range of methods included under the umbrella of AI auditing (figure 7).

Volunteer Audits

One of the auditor's major tasks is to design the different inputs to the system. In many cases, however, auditors

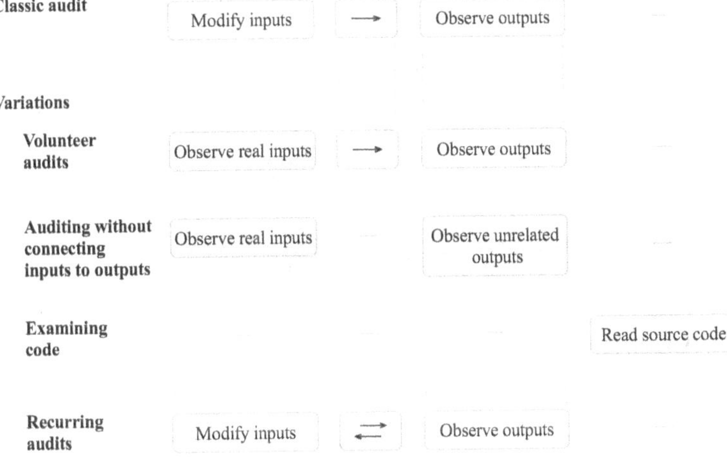

Figure 7 The "classic" AI audit design and common variations. Illustration credit: Shannon Yeung.

have no way to test carefully selected inputs. For example, in mortgage lending, it is illegal to include test or simulated data in an audit of a lending system. In addition, systems could be too fragile or important to insert test data. The auditors also might simply not have access to or the ability to control the system's inputs. So, what can they do?

In one example, auditors were interested in studying the political ads received by Facebook users during an election season.[23] But there was a problem: Facebook did not want them to. How could the researchers get access to ads run on Facebook if the company itself would not share any information or data with them? Their solution was to work directly with users, taking advantage of a feature that allowed users to download a list of the specific ads they had seen. By asking users to document and share their history of Facebook ads, the auditors were able to collect a large sample of political ads run on Facebook, with the added benefit that these were ads actually seen by a group of real users about whom the investigators had some added information, such as their demographics. Analyzing this data, they found that Facebook was not consistently enforcing some of its own policies—for example, it wrongly blocked ads for community centers, government websites, and civic events, mistakenly labeling them as political ads. The impact of these errors could extend well beyond the online space, weakening the ability of residents to connect with and participate in public life.

As we described in step 3, collecting data, this type of auditing has been called crowdsourced auditing. It often relies on volunteer users collaborating with an outside audit team to provide data where otherwise none would be available.

Beyond AI Experts: Users as Auditors

Auditors rely on user-provided data to answer questions in the crowdsourced audits we just described. We see great promise in this approach and several of us have been examining the potential for users to play a more prominent role in AI auditing.[24]

Users often see recurring issues with AI systems, including ones that expert auditors might not notice. In the case of Latanya Sweeney's audit described earlier in the chapter, searching her own name led her to notice problematic biases in the ads at the top of Google Search result pages. But not all users are professional computer scientists like Sweeney. Recent developments in auditing seek to bridge this gap, providing users with tools built specifically to enable them to ask and rigorously answer auditing questions.[25] These new tools confirm that such audits can identify issues that expert auditors might not think to test for.

As users increasingly interact with AI systems through social media, search engines, LLMs, and other user-facing tools, we see more potential for them to be involved in AI audits. One way may be to contribute data, but users might also contribute their ability to use specialized tools specific to the target domain to ask and answer new and interesting questions based on their own experiences. We will discuss potential paths to achieve this in chapter 6.

Auditing Without Connecting Inputs to Outputs

In some cases, auditors cannot access the AI system's inputs *or* outputs. They may be unable to test the system directly with different inputs, or access outputs through crowdsourcing or other means. In these cases, auditors might instead deploy a variation of the classic audit that analyzes whatever attributes of the system are visible to the auditors. This variation does not use repeated test data. We saw one such example at the start of this book when journalists identified illegal selection criteria in Facebook's ad targeting system that allowed advertisers to illegally discriminate against racial groups, even for housing and employment ads. At the time it was not known whether or how the auditors could collect the actual ads seen by a large group of users and conduct a classic audit proving that such racial discrimination had occurred. Instead, the journalists showed that the design of the ad placement platform made it easy to create illegal advertising. The audit visually inspected Facebook's ad posting website and attempted to post advertisements that should have been illegal.[26]

AI audits where inputs and outputs cannot be connected are usually less powerful than those following the more classic structure we laid out previously since they cannot assign responsibility as precisely. In some cases, though, as with Facebook's ad delivery system, the stakes and legal penalties are serious enough to lead to meaningful change.

Examining Code

The classic auditing method we outlined earlier in this chapter has developed largely because (as described in chapter

2) early auditors generally only have outsider access to the system and cannot see or analyze the code behind it. Ironically, sometimes auditors have the opposite problem.

In some cases, auditors cannot interact with the deployed system, its outputs, or its data. Instead, they may *only* have access to its software code. This can happen in a situation where a system's source code is required to be made public by regulation or a contract. In this scenario, an audit might take the form of reading the software's source code and determining what it might do wrong when deployed in practice. This is a common structure for security testing and is well known as a way to investigate voting machines.[27]

All the above variations present unique challenges that will be explored later. An AI system audit ideally involves data, software, inputs, and outputs. The more auditors know about each of these pieces, the more likely they are to accurately characterize the system's behavior.

Recurring Audits

AI audit reports characterize a system at a particular point in time. Since many AI systems learn from new data as time passes, their performance can change after an audit has been completed. AI system behaviors also regularly change due to periodic updates from their developers.[28]

Fans of Justin Bieber experienced the pace of change in AI firsthand in May 2010 after Bieber dominated the list of trending topics on Twitter for several weeks. Then one day—much to the outrage of "Beliebers," as his fans are known—his name suddenly disappeared from the list. Twitter announced that it had deployed a new AI system that prioritized sudden spikes in interest, unlike the previous one that prioritized overall popularity. Bieber was still just as popular, but Twitter's definition of "trending" had changed.

This announcement from Twitter was a startling reminder that software systems are constantly changing. Just as your computer periodically reminds you to run a software update, other AI systems also update and change—but usually without your permission. Imagine that Twitter had published an AI audit of its trending system in April 2010, just one month before the change. Within a few weeks, the audit would have been obsolete, and anyone relying on the findings would have been misled. For Bieber, inaccurate information about the AI system could materially affect his business—studies at the time found that music artists who appeared in trends on Twitter could experience a thousandfold increase in album sales in the days following.[29]

Regular audits are an important way to keep track of changes in the behavior of an AI system. That might mean a new audit for every major system change, depending on the stakes involved. While recurring audits might seem like a burden, it pays to plan for them. When organizations expect to do them regularly, it creates incentives to turn audits into routine work that can be done in a streamlined and cost-effective way.

Governments now employ a variety of AI systems to carry out public services. Local and national governments purchase gunshot detection systems, unmanned drone systems, automated license plate reading systems, language translation systems, and many more. A popular desire to improve AI and forestall problems has led government agencies to mandate a growing list of documentation and procedures, including AI audits. While some mandated audits are sparse, other jurisdictions require very detailed information.

The need for routine audits is gaining recognition in state and city governments. In January 2023, the city of San Jose, California created an inventory of AI systems to catalog how the city used AI to interact with its residents.[30] This repository included an impact assessment for every approved AI system in the city and an AI fact sheet that described data, bias concerns, intended domains, and more. Essentially, city officials created a framework for managing citywide AI systems including a proposal stage, data usage protocols, public comment stage, final approval, and continued monitoring. The monitoring protocol requires reporting of annual metrics. As governments

If the facial recognition software used by police in Detroit had been built with racial bias in mind, Robert Williams may never have been wrongly identified and arrested.

are increasingly making use of AI tools, these are the sorts of protocols they need in place.

* * *

If the facial recognition software used by police in Detroit had been built with racial bias in mind, Robert Williams may never have been wrongly identified and arrested for a crime that he did not commit. Using examples from well-known audits like the Gender Shades audit and an extensive discussion of an audit of LLMs used to create resumes, this chapter introduced the steps of an audit. The classic AI audit design consists of varying the inputs to a system and examining the outputs, but there are also several useful variations on this structure, including volunteer audits, user audits, code audits, and repeated audits. Strong audits can lead to accountability. But what makes an audit strong? In the next chapter, we turn our attention to the question of audit quality and explain how to distinguish between good and bad.

4

INTERPRETING AUDIT RESULTS

It was late 2021 when Derek Mobley received his one hundredth job rejection. Mobley had applied to a wide range of jobs in the finance sector for which he was qualified, only to be rejected again and again. All the jobs he applied to had run their application processes on Workday, an AI system offered by a workplace and HR management company that regularly appears on Forbes' Fortune 100 list.

Mobley, who is Black and has a disability, began to wonder if Workday's system might contain a racial or disability bias that was preventing him from getting job interviews. He filed a lawsuit against the company in his home state of California. Within a few months, the US Equal Employment Opportunity Commission filed a supportive brief in the lawsuit urging a federal judge to allow the case to proceed.[1]

Across the country in New York City, Tanmay Manohar, an HR executive at the film production company

Paramount, was on the other side of the hiring equation. Charged with making Paramount's hiring processes more systematic and fair, Manohar signed on with Harver, an AI company that purports to help employers reduce total hiring time while also rendering "blockbuster" results on diversity. Before long, according to Harver and Paramount, their algorithm more than doubled the company's rate of hiring ethnically diverse candidates.[2] Those people stayed with the company longer than previous hires, and the algorithm shortened recruitment time by a third.

But then, for Paramount, a new challenge surfaced. In November 2021, New York City passed Local Law 144 that required anyone using a hiring algorithm to have that technology audited for racial and gender bias.[3] To comply with the law, Paramount commissioned an AI auditing firm. Once the audit was complete, the results were published in compliance with Local Law 144, allowing job seekers and regulators alike to get a closer look at Harver's tool. Paramount HR and the AI auditors knew that this was where the real work would begin: Their hiring tool would be subject to public scrutiny once the audit was published.

For Manohar and HR executives at hundreds of other companies across the city, the question became: How do you interpret an AI audit if you are unfamiliar with AI? Manohar suddenly needed to be an intelligent reader of the forthcoming audit report. And the same could be said for job seekers. Back in California, Derek Mobley was grappling with an AI system that had not been publicly audited. For job seekers in New York facing situations similar to Mobley's, the new law promised to open a doorway to understanding more about how these systems work.

Earlier chapters in this book introduced AI audits and their typical processes. However, to have an impact, people like Manohar and Mobley must understand audit results before they can act on them. To ensure that AI is safe and accurate, we need auditors. But it is also important that we develop the critical skills needed to read AI audit results. Every AI audit is only as good as its ability to be understood by an informed reader.

While we have focused on AI systems that may have problems and ways AI audits can help us understand those problems, AI audits can also be problems. As audits become widespread, we may see more audits with major flaws. This section will review the most common characteristics of a bad audit to help you learn to scrutinize these audits in an informed manner.

Why Audits Go Wrong

There are strong incentives for AI system owners to prefer bad audits if they do not fully understand how to integrate audits into their organization. AI audits are designed to detect a system's flaws—as such, they may surface problems

Every AI audit is only as good as its ability to be understood by an informed reader.

that can be expensive to fix, cause bad press, and leave companies vulnerable to legal challenges. A misinformed or nefarious system owner wants an AI auditing process reduced to a formality or a rubber stamp so that it is much less likely to reveal deficiencies in the AI system. Experts have called this *audit-washing*, borrowing from "whitewashing."

Another cause of low-quality audits comes from the pace at which the field is moving. The industry-wide race to build and deploy AI has prompted all kinds of people to brand themselves as AI ethics experts and AI auditors. AI audits, just like AI itself, can be a complicated domain to master, and the expertise of people who call themselves auditors varies widely. There is no widely accepted certification or standardized training for auditors, though a small cottage industry is emerging that purports to sell such accreditation.

A third issue comes from conflicts of interest. Even some audits by independent academic researchers have drawn criticism because of the potential for undue influence by the AI company under scrutiny. Even the appearance of a conflict of interest can taint an otherwise strong audit.

Evaluating an AI Audit's Framing

As we saw in chapter 3, an AI auditor has to make many important decisions before conducting the audit. These include posing a question to be investigated, deciding how to measure its key components, and selecting the threshold on which the system will pass or fail. Bigger picture questions also need to be considered. Have the auditors described how those decisions impact people? Have they identified the important institutions and organizations that play a role? Have they identified external forces or incentives that may be difficult to change? We grapple with each of these questions in the sections below.

Does the Audit Ask the Right Question?

All AI audits start with a question. If an AI system is built to label common objects in different images, a simple audit for accuracy might ask, does the system label images correctly? Sometimes these systems can confuse things that look similar in a certain way, even if they are entirely different. For example, a Chihuahua is a small, spunky dog breed, while a blueberry muffin is a breakfast treat. Most humans have no trouble telling them apart. But the humorous and frequent tendency for image classification systems to confuse the two has become a meme among AI engineers known as the muffin–Chihuahua problem.[4] This phrase can metaphorically refer to any situation where AI has well-known or expected weaknesses.

When an audit needs to test for bias, the task becomes more challenging. An auditor of an AI hiring system might ask: "Does my hiring system exhibit gender bias?" Yet it

The muffin–Chihuahua problem can metaphorically refer to any situation where AI has well-known or expected weaknesses.

is not at all obvious what bias means. The definitions of "muffin" and "Chihuahua" are straightforward, but bias audits must operationalize abstract ideas like "biased" and "fair." There is usually no agreement about a quantitative definition. In fact, researchers have shown that it is mathematically impossible to satisfy all definitions at once.[5]

In the case of the SABRE airline reservation system that we described at the beginning of this book, "display bias" (an obscure legal term from the airline industry) was a problem identified by smaller airlines like Braniff. They argued that American Airlines, the owner of the SABRE system, unfairly listed American flights first in search results even when it was not the best option for travelers. After extensive litigation and regulation, airline industry regulators in the United States settled on a definition of display bias in which any display of the results that does not literally implement the sort order that is written on the screen, such as "lowest price," is unfair and deceptive and counts as illegal display bias.

We could then imagine an audit implementing this rule and checking airline reservation sites for display bias. The audit question based on the definition of that term would be: "Does the sort order of the flight results match the name for the sort order displayed on the screen?" Unfortunately, airline reservation sites responded to this definition of bias by changing the *name* of the sort order displayed on the screen instead of ceasing to manipulate carrier and flight rankings in opaque ways. Rather than stopping practices like promoting or demoting some carriers to serve their business interests, airline reservation systems have come up with new words for their sort order, such as "relevance" or "best."

As a result, online airline reservation systems today would pass an audit for display bias; under that definition, no bias would be found. In other words, current systems could satisfy a check for display bias because they have changed their interfaces to avoid sorting by concrete, objective criteria like "shortest trip" or "lowest price." If an airline reservation system is recommending itineraries because it makes the airline more money, as long as the displayed sort order is labeled "recommended," this technically does not exhibit display bias.

This display bias loophole highlights the importance of attending to the specific question an audit asks. Informed audit readers will ensure the audit's central question is strong enough to prevent such loopholes, especially those created through excessively forgiving definitions of bias that allow AI systems to follow the letter of the law rather than its intent.

What Questions Does the Audit Not Ask?

Along the same lines, readers should consider the questions the audit does not ask in addition to the questions it does. An audit might be rigorous and convincing, but as an audit report reader, you might entirely disagree with its focus or goals.

For example, New York's Local Law 144 is only concerned with identifying the biases in automated hiring systems that violate local and federal law. Audits under Local Law 144 produce a selection ratio of one category (e.g., women) compared to the pool of all applicants, and only consider specific types of bias (i.e., gender and racial bias). However, these audits are not required to test for other types of discrimination, nor do they evaluate the system's accuracy in assessing applicant qualifications or matching the best people for a given job—things most people evaluating an automated hiring system would care about.[6]

Whether you are an employer or a job seeker, as a reader of a Local Law 144 audit, you may find the information you are given lacking, and your evaluation of the audit might be to conclude that it focused on the wrong question altogether.

What Threshold Does the Audit Use?

Suppose you are interested in possible discrimination due to automated hiring tools and set out to read a Local Law 144 audit (as we will later in this chapter). You may be satisfied that the audit looks for differences in hiring scores between men and women applying for the same jobs, but what difference between the two hiring rates would you consider to be a problem?

A common rule of thumb used in AI audits about hiring is a threshold set by the federal government: Keep the ratio of the hiring rates above four-fifths (80 percent). If your hiring (or selection) rate for any group is less than 80 percent of your selection rate for the group with the highest rate, that can be considered evidence of adverse impact on the disfavored group in violation of federal antidiscrimination law.[7] That is, if women are less likely to be hired than men, and equal numbers of men and women apply for a job, at least four women should be employed for every five men.

By choosing this criterion, the auditor is essentially saying that differences smaller than four-fifths are not a problem—but many people might disagree. Even the federal government might not agree, since the four-fifths rule is merely a guideline.

Many early AI audits tested for racial discrimination by checking to see if a system violated the four-fifths rule for a set of racial categories defined in the laws on employment discrimination. Importantly, current experts on employment discrimination often feel that the four-fifths threshold, and even the racial categories that are widely used, are no longer useful. Before AI, decades of work to combat employment discrimination using these criteria failed to halt discrimination, which was the original purpose of these rules.[8] Nonetheless, these thresholds and categories have been moved from their origin in offline employment audits into AI audits. They have moved to housing, health, social media, and the legal system. There was no agreement that these thresholds would work or were a good idea. Rather, they were the only thresholds that were easily testable, and there was at least some precedent for using them.

While you might agree with the question that motivates a given AI audit, you still need to pay attention to the standard or criterion against which results will be measured. Remember that the threshold may have been chosen simply because it has been used before. It might not be particularly applicable to your context, and there may not be any correspondence between the threshold and the actual harms that you would like the audit to detect or prevent. Instead, look for evidence in the audit that the threshold used relates to any available real measures of harm.

Does the Audit Compare Meaningful Categories?

In the examples we have seen above, the categories being compared (different gender or racial identities) in the context of hiring are clearly meaningful. Although they are nuanced—they are not fixed, objective, or binary—there is a connection between these identity characteristics and hiring patterns in our society. However, the definition and connection of the categories being compared in an audit are sometimes unclear.

When presented with a set of categories, you should view them critically and draw on the expertise of those who understand details of the domain where they are applied. Any categorization is a simplified representation of the real world, so asking how closely these categories hew to the ideas or people they are meant to represent is an important step. It might be possible to argue with the quality of any categorization. Back to the muffin–Chihuahua problem: It is relatively easy to assign categories to house pets and baked goods. It is more challenging to assign categories that aim to reflect things like a person's ethnic identity or personality type.

Ensuring that the categories being compared in an audit are well understood and have plausible and important connections to the AI system's function is good practice. Verifying that they are meaningfully linked to the real world is imperative.

Evaluating an AI Audit's Execution

Once you are convinced that the audit is correctly structured—that it asks important questions, measures them using meaningful constructs, and evaluates against the right standards—the next step is to evaluate its execution.

Who Is Excluded from the Data You Are Using to Test?

Many datasets intended to represent a certain group of people have significant gaps, even by design. New York

City Local Law 144 allows independent auditors to exclude groups that represent less than 2 percent of the individuals in the dataset from analysis. Since some employers receive hundreds of thousands of applications, excluding 2 percent could mean excluding thousands of people.

We can also imagine situations where an important problem with a system under scrutiny causes people to stop using it. If some aspect of a hiring tool is so offensive that it prompts people in one group of applicants (say, one racial or ethnic group) to disengage from the system, an audit might not be able to detect the problem or might misdiagnose it.

A recent Pymetrics AI hiring system audit showed that the company used self-reported demographic information collected at the end of the applicant's engagement with the system. This might seem reasonable at first glance, but that choice meant that the company could only consider the demographics of candidates who had successfully completed the system's entire process. This could mean that applicants who were uncomfortable answering demographic questions—or anyone whose computer crashed before the end—would not be represented in the audit data.

Another recent audit investigated HackerRank, a popular product that automates technical interviews for software engineering positions.[9] One of HackerRank's tools is a system that attempts to detect cheating and plagiarism by taking photographs of applicants during the test to see if they leave at any time or if someone else takes the assessment. Under Local Law 144, HackerRank was required to conduct a bias audit of this system. To test the system based on gender, race, and ethnicity, as required by the law, the auditors used a publicly shared dataset of faces called FairFace, where people's photos were labeled to indicate their gender, race, and ethnicity.

But the people whose faces appeared in the FairFace dataset did not have the chance to label their own faces according to their race, ethnicity, and gender. Instead, they were categorized by online crowd workers, who are typically paid pennies per label.[10] The designers of FairFace asked three to six randomly chosen workers to label each image with one of seven racial categories. If the labelers disagreed on the categorization of the person in the photo, it was discarded from the dataset. This meant two things: Each image was labeled with a single racial identification, and the dataset only included people whose identification the labelers agreed on. So, by definition, there are no photos in the dataset that the labelers deemed racially ambiguous. (More than 10 percent of US citizens identify as mixed race.) And if a racial category had too few labeled images—as happened for Pacific Islanders—the whole category was omitted.

The numerous shortcomings of the FairFace dataset meant that the audit could not draw conclusions about

Pacific Islanders or anyone whose racial appearance was not obvious to randomly chosen strangers.

As auditors, we always evaluate the data sources used in an AI audit for completeness as a key part of data quality. As a reader, asking yourself where the data came from, as well as who it includes and who it excludes, is a critical step.

How Fresh Is the Data?

It might seem obvious, but if you lack confidence in the data used for an audit, you should not trust the audit results; a knowing reader of an AI audit examines the sources of an audit's data and considers its quality.

Even when the data used in the audit adequately captures all the relevant groups, it may still have flaws. One aspect to attend to is the freshness or recency of the data. AI systems are typically trained using pre-existing datasets that help the model learn whatever task it is meant to perform. However, many experts have pointed out that training data are inherently retrospective: By definition any existing data can only be drawn from the past, producing systems that act based on what has been seen before. Although obtaining the freshest data possible solves some problems in both AI systems and audits of them, some problems related to freshness may not be solvable.

This can be a major problem complicating AI. For instance, Amazon's AI system for reviewing resumes was intended to make hiring more efficient. But HR leaders needed to be sure this efficiency did not come at the expense of candidate quality. Informed by the biases they knew existed in their past hiring decisions, they conducted an internal AI audit to determine whether the system unfairly discriminated against female candidates for software engineering jobs.[11]

We do not have the details of the internal Amazon discussion, but we know that in the company's history, women were less likely to be hired than men—this was the problem the company was trying to address. The hiring system was based on ten years of historical hiring data, and when studied, Amazon developers also found that the system preferred candidates who used masculine-coded language like "executed" or "captured" on their resume. The audit revealed that Amazon's system reproduced the same biases found in the historical data that was used to build it. Upon discovering this, Amazon was able to make an informed decision: They stopped using the tool altogether.

Careful audit readers should be aware: Whenever auditors evaluate a system using the same biased historical data that trained their model, the potential for reproducing harmful historical problems is high.

How Was Sensitive Information Obtained?

Many audits require auditors to obtain and handle sensitive information about real people. When embarking on an antidiscrimination audit, for example, auditors need to analyze users' race, gender, and ethnicity. But this is easier said than done. Some industries have rules or best practices prohibiting them from collecting demographic information. These regulations and practices exist to protect people's privacy and reduce the risk that they will be subject to discrimination based on these factors.

Paradoxically, measures intended to protect people from discrimination can also make it more difficult to fight discrimination. Without access to these sensitive demographics, measuring discrimination against these groups is impossible. In some industries, such as mortgage lending, companies were previously forbidden from collecting sensitive data like racial or ethnic background. This approach was later replaced by laws *requiring* companies to collect data about race, which now coexist with rules prohibiting mortgage lenders from discriminating based on race.

After the passage of Local Law 144 in New York City, the auditing firm BABL AI was faced with the prospect of producing AI audits about discrimination for employers and sought clarification from the city government. How did the city expect the audits to work? How would it handle situations where companies did not collect or retain this sensitive personal information? In cases where companies did not have reliable records, could auditors guess the demographics of past applicants? Or would it be better to test the AI system using simulated job seekers of different identities?

The challenges faced by BABL AI are routine hurdles faced by any organization trying to analyze sensitive information. Auditors might need to negotiate data security agreements with their clients and agree on how much information about the underlying AI system the auditor is allowed to see. Interviews with more than a dozen auditors indicate that these arrangements can be complicated because the users or operators of AI systems are not typically their creators. Most often, operators have purchased the systems from a third party. The amount of information the vendor provides its clients can vary substantially. A successful audit needs information from the AI vendor, operator, auditor, and users.[12] When reading an audit, a savvy reader will check these relationships for potential issues.

Although this section foregrounded how auditors handle sensitive personal data using a focus on demographics, the same situation holds for other secret, sensitive, or proprietary details. Consider the case where the internal operation of an AI system is not known because the auditor is an outsider. It is common for an outside audit to result in accusations from the AI system's operator that the auditors simply misunderstood the AI system because they did not have full access to it. This may be the case, but an external audit can still make claims adequately scoped to the findings they are able to make without internal access. A defensible audits states: This is how the system behaved when it was tested.

Audit report readers should always check how sensitive data was obtained and look for disclosures about simulated or approximated data. At a higher level, to ensure that the audit draws reasonable and evidence-based conclusions, readers should ask how auditors arrived at answers to questions that require secret or sensitive details.

Audit Interpretation in Practice: Two Examples

Now that you've been introduced to the aspects to consider when reading an audit, we will work through two examples. Both are drawn from documents produced to comply with Local Law 144. While the law requires public notice and disclosure, it does not aggregate or publish the reports on a government website. Nevertheless, nonprofits and nongovernmental organizations like the ACLU have produced public archives of these documents.[13]

Paramount's Use of Pymetrics

Perhaps better known as the producer of *Spongebob Squarepants* and *Star Trek*, the Paramount conglomerate is also a major employer, one that has contracted with AI hiring companies like Harver (as you may recall from the start of this chapter) and Pymetrics. To comply with New York City's Local Law 144, Paramount needed to commission bias audits of both Harver and Pymetrics, specifically

Bias Audit for New York City Local Law 144
Prepared by BABL AI Inc. | 06/29/2023
Letter from the Lead Auditor | Summary | Conclusions | Findings

Q.B. Testing dataset: The dataset on which disparate impact was quantified shall be defined and characterized. Q.B.1. Evidence shall show justification for why the selected dataset was appropriate for disparate impact testing. Q.B.2. Where test data as defined in § 5-300 was used, evidence shall show a. justification for not using historical data, b. that historical data is not sufficient to perform a statistically significant disparate impact testing, and c. the methodology by which test data was collected Q.B.3. Where disparate impact testing was not completed by BABL, evidence shall show a. that the most recent testing was conducted less than one year prior to the start date of this audit, or after a major update to the model, unless the update was more than one year prior to the start date of this audit, in which case, evidence shall show b. justification for why such testing was still appropriate. Q.B.4. Evidence shall show that the data used in the testing was within one year of the start date of the disparate impact testing.	**PASS**

Testing conducted by: pymetrics
Date of last testing: Apr 13, 2023
Time span of data: Jan 01, 2022 – Dec 31, 2022

From pymetrics: The dataset contains historical application data from "all models used to evaluate candidates between January 1, 2022 and December 31, 2022." Candidate demographic information was "collected via a voluntary exit screen at the end of the pymetrics assessment".

Figure 8 An excerpt from the BABL AI audit of Paramount's use of Pymetrics, showing a description of the dataset. Although this audit looks official, the wording obscures important information.

regarding its own use of those companies' services. Its audit focusing on Pymetrics, conducted by BABL AI, is excerpted above in figure 8.[14]

On the surface, the report looks impressive, offering twenty-one pages of policy details and tables of results that suggest the system does not display illegal gender or racial biases. But the report excludes or obscures information about many of the crucial questions we have addressed in this chapter, such as how fresh the data is and what demographic groups were left out of the audit. We briefly describe these two shortcomings and how they can be discerned from the audit report below.

How Fresh Was the Data?

The audit report emphasizes one conclusion: Paramount's use of Pymetrics earns a "PASS." Only when digging into the report's specifics does the reader find important details about the data, like when and (briefly) how it was collected. Was the data fresh? The audit reports the data collection dates as between January 1 and December 31, 2022. An astute reader will note, then, that this audit only gives us insight into the system's behavior in 2022, not before or since.

Who Was Left Out?

The description of data collection has a notable red flag: Demographic information was "collected via a voluntary exit screen at the end of the assessment." As described above, voluntary data collection on an exit screen will be incomplete, because some people will decline to share such information about themselves and others may exit before that step. This section does not reveal how many people volunteered their demographic information. But a later section (buried under five tables of unrelated information) tells us that more than 200,000 applicants did not report their gender or race/ethnicity. This sounds like a large number, but it might not be an issue if millions of people apply for jobs at Paramount.

Because the audit does not directly report the percentage of applications that lacked self-reported demographics, and it does not report the total number of applications, a highly motivated reader might calculate the total number of applications themselves from the data available in the tables provided. If we add together the number of applicants who selected male (197,821) and the number of applicants who selected female (250,692), we find that 448,513 people did report their gender. This means that the number of people who did not report their gender or race/ethnicity is actually a substantial fraction of the total applications: 30 percent of the users of this system are missing.

Should we trust an audit that can only evaluate the experience of seven out of ten people using the system? We have no way of knowing if the other three people were harmed by bias in the system. This question is one each reader might answer differently depending on the context and its stakes. Still, a more trustworthy report would have been transparent about these facts and presented them directly.

In Paramount's case, the available data suggests that the rates at which applicants from different groups are selected for hire are relatively equal. This is good news for the company and a reader concerned about discrimination. But if audits only report the minimum information required by law, most audits will only be as good as the law that mandates them. And if the law does not require important information to be assessed, the audit may not cover that at all.

The major lesson from reading this report is that key pieces of information—like the number of applicants with missing data—may be obscured, or their relative importance may be hard to ascertain. A very motivated reader will need to make calculations or decisions themself. When descriptions of criteria or data reveal issues or when those descriptions are missing, it is safer to be skeptical.

NBCUniversal's Use of SmartRecruiters SmartAssistant

A different example from NBCUniversal illustrates the potential benefit of audits. NBCUniversal hired a different auditor, ConductorAI. On the surface, this audit appears much less impressive. It fits on a single page with no details about policy and only a very focused set of results.[15] Again, the appearance is deceiving. Unlike the last audit, this one includes many crucial facts BABL AI omitted.

The report begins with a summary of the findings (figure 9). The data are quite similar: They also use real historical data from the system, which was collected over the previous year. But in this case, the report describes the total number of applicants and distinguishes between applicants who chose not to provide gender information, race/ethnicity information, or both. And, even better, this additional information informs the reader that only 2.5 percent of applicants provided no demographic information. If we compare this with the 30 percent from the previous example, we can feel much more confident in these results.

In addition, the report includes impact ratios for all categories of applicants (figure 10). Even though ConductorAI is not required by Local Law 144 to report values for race/ethnicity categories that are less than 2 percent of their applicants, it includes them for all groups. The reader does not need to calculate these values, making it much easier to see that no group has an impact ratio that should be a concern.

Comparing these two audits is a good example of the advice "do not judge a book by its cover." While the long documentation of BABL AI's audit looks impressive and official, the simple one-page audit carried out by ConductorAI provides more information about the data and

Summary of Findings:

The audit results shown below were produced on 07/11/2023.

The audit was performed on data from the following tool: SmartAssistant.

The audit below was performed using historical data from between 07/01/2022 and 07/07/2023 gathered from multiple employers provided by the AEDT vendor to the independent auditor, ConductorAI.

SmartAssistant produces scores that are not directly comparable across jobs. As a result, the initial scores have been normalized across all job types to produce final scores that are comparable across the entire dataset for the purposes of this audit. This normalization process created a final score for a candidate by multiplying their initial score by the median score of the dataset divided by the median score of their job type.

It is the independent auditor's opinion that the provided data is representative, and "Scoring Rate" is most appropriate for assessment. Scoring Rate is the percentage of candidates within a given group whose score is above the median score of the entire dataset. This median score across all candidates is 66.0. Impact Ratio is calculated by dividing the Scoring Rate of a given group by the highest Scoring Rate within the category.

The dataset given to ConductorAI consisted of 4,196,161 candidates. For 105,724 of these candidates, no sex or race data was known, so they are excluded from all categories below. 162,872 candidates had sex data but unknown race data. 19,747 candidates had race data but unknown sex data.

Figure 9 An excerpt from the ConductorAI audit of NBCUniversal's use of SmartRecruiters SmartAssistant. It features a comprehensive dataset description.

Race/Ethnicity Categories	# Of Applicants	Scoring Rate	Impact Ratio
Hispanic or Latino	470,904	45%	0.89
White	1,457,444	46%	0.9
Black or African American	796,447	48%	0.96
Native Hawaiian or Pacific Islander	14,904	45%	0.9
Asian	941,554	50%	1.00
Native American or Alaska Native	29,612	43%	0.85
Two or More Races	216,700	46%	0.92

Figure 10 An excerpt from the ConductorAI audit of NBCUniversal's use of SmartRecruiters SmartAssistant. Impact ratios are reported for all groups, even those small enough that they would not be required.

findings of their audit. That data is more detailed and of better quality and helps anyone reviewing the report to be confident that the audit was done well and can be trusted.

You do not need to become an expert to recognize when an audit is transparent or if aspects are left out or hidden. And comparing different audit reports can be an excellent way to strengthen your sense of whether a system really meets the relevant standards.

* * *

This chapter sought to develop the reader's taste for AI audits, explaining why one audit might be preferable over another. It described a set of common mistakes, introducing concepts like audit-washing and the muffin–Chihuahua problem. For instance, savvy readers of audits will be alert for stale data, excluded groups, missing questions, flimsy thresholds, and retrospective training data. New York's Local Law 144 has produced a cornucopia of AI audits, and this chapter closed with a comparison and evaluation of two audit reports about employment software. In the next chapter, we will discuss how audits can contribute to all these potential outcomes—improving products, building consensus, creating new policies—and the risks associated with each.

5

AFTER THE AUDIT

An AI audit in the hands of someone committed to the common good is a powerful tool for starting a conversation, prompting change, or defending a working product that has been wrongly accused.

Audits have provoked major governmental action and sparked public debate on important questions about fairness and equality in the age of automation. They have saved some companies millions of dollars and forced others to change course due to public or regulatory pressure. They have kept the innocent out of jail, prolonged lives, and led to the reform, redesign, and improvement of many systems.

Outcomes: The Broadest View

Recall the 1970s SABRE airline reservation system discussed at the start of this book. Audits determined that

American Airlines, the system's owner, prioritized its flights on the new computerized reservation system designed to serve the whole industry. What happened next? Government regulators took action. The Civil Aeronautics Board introduced regulations prohibiting American and other airlines from rigging the display of available flights. The board further imposed fines on entities that violated these regulations. It mandated the creation of a feedback process for a broader audience of interested parties. It passed a transparency rule to reveal SABRE's criteria for ranking flight search results. These actions can be directly traced to the outcomes of the SABRE audits conducted by American Airlines' competitors.

The SABRE audits also opened a public conversation about values. How should the system work? What counts as fair competition? How should regulators step in and address the issues at hand? American Airlines argued that because it had made major investments in SABRE, it deserved to reap the financial benefits of rigging the system, and its actions were perfectly acceptable and even desirable. If American's competitors didn't like SABRE's listings, they could build their own airline reservation system and manipulate it as they wished.

American's argument did not prevail. However, the mere process of making the argument—through hearings, testimony, and legal filings—revealed a great deal of information about the SABRE system that was not previously public knowledge. The event became a milestone in the development of the airline industry. It informed subsequent regulations and debates far beyond air travel, such as whether it is fair for a cable or satellite television provider to privilege some channels over others on menu screens, and the debates about AI described in this book.[1] The SABRE audits resulted in regulations, but they also led to press coverage and public disclosure that fed into broader national narratives about fairness and competition.

Public audits that lead to debates are the first visible green shoots of auditing that sprout above the ground. But quieter AI audits are conducted all the time. Largely unseen efforts routinely help companies test and answer questions about their products, assess whether they are vulnerable to regulatory action, and evaluate their potential for existential reputational harm. In this chapter, we consider what comes after an audit, demonstrating how audits can improve products, trigger new regulations, and inform the public at large.

Audits Improve Products and Reduce Liability

Companies commission audits to refine their products, prevent bad press and consumer backlash, and ensure ongoing compliance with legal obligations. A good audit offers a competitive advantage and may be a key part of a company's business strategy.

AI audits frequently ask, does this system do what it claims to do? If a company audits its own system in advance of deployment and releases the results, potential customers can be assured that a system is fit for their purposes and that they are acting responsibly by choosing it. If an audit reveals bias or poor performance that was previously not known, this can protect a company against reputational harm and liability if the problems are corrected. A well-audited system, particularly one audited by an independent third party, may become a preferred vendor for an entire industry—especially in a heavily regulated industry where the high stakes of the system can make customers afraid to adopt new, untested AI. This can become a form of marketing for the company if it addresses the problems the audit finds.

Audits Pave the Way for Product Improvements

When an audit elucidates a problem with an AI system, it creates an important opportunity for product improvement. Word embeddings—computer models that represent the relationships between words numerically so that AI can work with them—are a key ingredient in AI tools for speech recognition, grammar checkers, translation software, and large language models. Bias within word embeddings is a significant issue in many applications of AI and can impact the quality of products like Google Translate that use AI to automatically translate text between different languages. For instance, when translating from Turkish, a language that does not contain gendered pronouns, to English, a language with gendered pronouns, Google Translate has defaulted to masculine pronouns when referencing doctors, and feminine ones when referencing nurses.[2] Auditing can identify these problems. An audit of the w2vNEWS word embedding found gender stereotypes in the representation of occupations (e.g., medical occupations such as doctor were associated with men, while nurse were associated with women). To address this, developers produced a method for modifying the system to mitigate those stereotypes.[3]

Quality audits that improve software are a worthwhile approach that can help a company develop a competitive advantage, but there can be challenges along the way. In this section, we explore some of the ways that companies have improved their technology following audits, and some of the challenges they've encountered along the way.

Uncover Underlying Causes

Why did the AI systems in Joy Buolamwini and Timnit Gebru's Gender Shades audit have such a hard time accurately labeling the faces of darker-skinned women? Systems like these are typically trained using thousands of photos of real people—over time, the computer model "learns" what a person looks like, based on patterns that it identifies in those images.[4] Buolamwini and Gebru suspected that the images used to train the models for these systems did not include nearly enough dark-skinned faces, and the systems were therefore unable to properly label them.

The Gender Shades audit served as a wake-up call for the companies in question, prompting them to revisit their procedures for developing training data. Soon, they were collecting a much more representative range of images. In a follow-up audit, their performance improved with an up to 30 percent reduction in error, reducing disparities between groups.[5]

Questions about underlying causes come up in many settings where audits can be useful. Imagine entering the phrase "Donald Trump" into X's search bar in 2024 (figure 11). The results might include tweets by Donald Trump and other tweets commenting on Donald Trump. Some tweets are supportive; others are not. Earlier audits of this platform found that the search results can be dominated by one side of a political issue. If a user receives a list of all-positive or all-negative search results, why is this happening?

A Twitter search can only draw from the available tweets. If a system like Twitter search (that highlights tweets in response to a user's query) leans toward one political party versus the other, it could do so simply because people in one party post more tweets. Auditors needed to discover how results came to be dominated by just one side

Audits of X found that the search results can be dominated by one side of a political issue. If a user receives a list of all-positive or all-negative search results, why is this happening?

Donald Trump

Using a child with cancer as a prop, while simultaneously cutting funding for the very research that keeps him alive, is quite possibly the most disgusting thing I've seen Donald Trump do.

220 1.6K 4.5K 93K

Donald Trump is whining that Democrats aren't standing, cheering, or clapping for him.

He's such an embarrassment.

Donald Trump claims that no one has ever heard of a country called Lesotho.

What you won't hear from Donald Trump tonight? A plan to improve the lives of working people.

Figure 11 In a search on X for the words "Donald Trump" conducted in March 2025, all the top results are critical of Trump. Many audits have asked what causes this to happen. (This screenshot has been redrawn for legibility with profile images removed.) Illustration credit: Shannon Yeung.

of a debate. In 2017, a team of auditors including two authors of this book (Motahhare Eslami and Karrie Karahalios) developed new auditing methods to expressly distinguish between two possible causes. They asked: Was one political view dominating the results because that view actually dominated the tweets people posted? Or was the Twitter search system putting a thumb on the scale, suppressing one viewpoint and promoting the other? In 2017, they found that Twitter's search algorithms were not biased toward Democrats—a concern some had voiced at the time. Instead, there was simply a larger number of overall tweets about the candidates that reflected Democrat-leaning values.[6] The search system was a mirror showing what actually existed, and it passed their audit for bias.

Band-Aid Solutions

It can be challenging to ensure that post-audit repair truly resolves the underlying problem, rather than merely masking it. In 2015, the Google Photos app incorrectly tagged a photo of two Black people, labeling them as gorillas. The company issued an apology, but it did not address the underlying problem. Instead, it chose to eliminate the "gorilla" label entirely, even for images that actually depicted gorillas.[7] It also removed "chimpanzee" and "monkey" as labels. Eight years after the fact, an audit by *New York Times* showed that neither Google's photo app nor Apple Photos could label gorillas.[8] It remains uncertain whether

Google Photos' AI system is still restricted due to a technical inability to solve the issue, a lack of resources, or excessive caution.

But a band-aid can be a good way to prevent further harm until a permanent solution is found. Margaret Mitchell, cofounder of Google's Ethical AI group at the time, advocated for removing "the gorilla label, at least for a while." She stated, "You have to think about how often someone needs to label a gorilla versus perpetuating harmful stereotypes . . . The benefits do not outweigh the potential harms of doing it wrong."[9]

Band-aid solutions risk masking underlying issues and preventing companies from identifying root causes and developing long-term solutions. But they can offer immediate action, particularly in high-stakes situations where the risks of inaction are significant if a permanent fix is not found. Nonetheless, it is important to ensure that these temporary measures are not retained in the long term, as this can hinder progress toward achieving lasting and meaningful improvements.

Audit Again

Since band-aid solutions offer a quick fix rather than a deep investigation of the problem, they can mean that complex problems—often ones where issues can arise from more than one source—can be missed. A good way to guard against this is to repeat audits.

After a 2016 ProPublica investigation proved that Facebook had enabled advertisers to discriminate against Black and Hispanic users on the basis of "ethnic affinity" groupings, the company vowed to step up its game.[10] Facebook responded by improving the ad tools, reporting in February 2017 that it had made improvements in "the areas of housing, employment, and credit, where certain groups historically have faced discrimination"—a plain reference to its obligations under the Fair Housing Act.[11]

Later that year, ProPublica researchers did what good auditors do: They ran a new audit.[12] Researchers purchased a series of rental housing ads and, with the tools in Facebook's advertising interface, they deliberately excluded several types of people from seeing the ads. The interface allowed them to exclude African Americans, mothers of high school kids, people interested in wheelchair ramps, Jewish people, expats from Argentina, and Spanish speakers. Each of these groups are protected by the Fair Housing Act. But within minutes, Facebook's ad system approved every single ad. Although Facebook had removed advertisers' abilities to target "ethnic affinity" groups, other kinds of groupings were still allowed, and this enabled the same illegal discrimination to occur.

The problem has continued to haunt the company, with Facebook settling two major lawsuits on the issue—one in Washington state and one at the federal level. In March 2019, now-former Facebook executive Sheryl Sandberg announced changes to housing, credit, and employment advertisements, specifically stating that it was no longer possible to target by age, gender, zip code, or multicultural affinity in those domains.[13] Unfortunately, with the removal of these categories, Facebook's ad delivery mechanism instead used only an opaque AI system to determine which audiences to show each advertisement to. Even if an advertiser chose to show its ads to an "untargeted" or generic audience, the AI system would show the ad more frequently to some categories of people than others.

These repeated audits by ProPublica and other researchers slowly narrowed and revealed new causes of Facebook's problematic ad targeting and delivery system.[14] And they showed over time that the fundamental issue—discrimination in ad delivery on the basis of protected categories like race and ethnicity—was occurring through many mechanisms. Different systems, algorithms, and data sources led to the same bias. When one source of bias was removed, others remained, or new ones appeared. The team at ProPublica was prepared for this and was right to carry out repeated and consistent investigations over time to address the problem.

Audits Provide Reputational Benefits and Protections

An audit can and should be an opportunity for good companies to attract good press that allows them to showcase both the performance and fairness of their systems. When the Pymetrics corporation found itself facing justified skepticism of its AI hiring and talent management product, the company paid outside experts to conduct an audit. Pymetrics later received widespread coverage that cast the company in a positive light, including headlines like "Pymetrics attacks discrimination in hiring," and "People are terrible judges of talent. Can algorithms do better?"[15]

Self-commissioned audits of AI systems like this one can build on existing strategies by corporations to protect against reputational harm. But whether self-commissioned or external, audits can provide a way for companies to showcase their AI systems' accuracy, fairness, and other capabilities. When a competitor gets bad publicity for an unfair system, it presents an opportunity to showcase superiority. In the case of the Gender Shades audit of face detection systems, the companies targeted by the audit used it as an opportunity to improve the accuracy of their AI systems.[16] The follow-up audit showed that they had succeeded in reducing error rates.[17] Strong audit results can lead to positive press and a competitive advantage.

Audits Inform Decisions to Abandon a System

An AI system's flaws are sometimes unfixable. Some audits reveal problems so severe that the risks and harms of a system might outweigh its intended benefits. In these cases, operators may choose to discontinue or abandon the system entirely.[18]

AI systems that peddle "predictive policing" services are one cautionary tale. Companies like PredPol have claimed their systems can predict when and where crimes are likely to take place and have sold their software to police departments across the country. But after internal and external audits laid bare their flaws—audits showed bias and plain old inaccuracy—the systems saw a sharp decline in use. Multiple cities, including Los Angeles and Chicago, have ended their predictive policing programs, due in part to these audits.[19]

While abandoning a system might seem like a failure, it can represent a positive decision. When organizations choose to discontinue an AI system, they acknowledge that the costs of maintaining a flawed system, both ethically and practically, outweigh the benefits. The decision to abandon a product enables leaders to focus their efforts and resources on more promising innovations. If an audit reveals unacceptable levels of bias or widespread inaccuracy, then a company can make changes in advance of deployment, which can protect it against reputational harm and potential legal liability.

Audits Enable Resistance and Circumvention

Audits can also be used in novel ways to advance public debate about the use and operation of AI. Inspired by concerns about the potential harms of computer vision–powered mass surveillance, artist Adam Harvey used AI audit techniques to create a fabric pattern called "HyperFace," containing an abstract series of squares that some AI systems would misidentify as a human face. When worn on clothing, the pattern draws these systems' attention away from the wearer's real human face by presenting the AI system with a larger and more compelling false alternative—the patterns of dots in figure 12.[20] Conceptually, if HyperFace gained widespread popularity it could pollute the data sources used by face detection systems. That is, if the HyperFace fabric filled with potential false positives became popular enough, training datasets could be filled with this dot pattern rather than real human faces.

HyperFace was a compelling and effective thought experiment rather than an attempt to solve the problem of automated surveillance, but it illustrates how audit techniques can be used by outside parties in ways beyond the standard template.

Adam Harvey was making a political point about the dangers of technology, but consider a related example: the computer vision systems used by self-driving cars. AI audits found vulnerabilities in these state-of-the-art systems, revealing that affixing stickers to a stop sign could prevent a car from "seeing" and stopping at the sign.[21] We can imagine a dangerous situation where a determined anarchist effectively turns off traffic control signals for self-driving cars with a few well-placed stickers, traffic chaos ensues, and pedestrians and passengers are injured

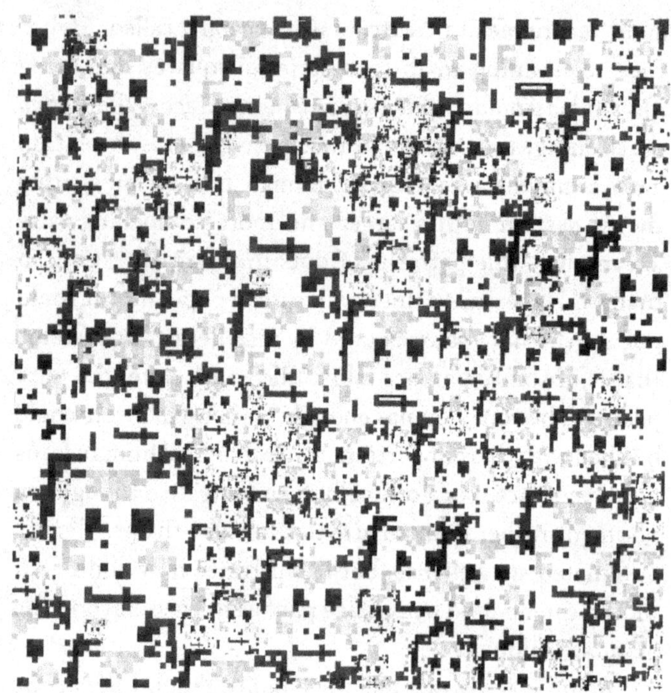

Figure 12 The HyperFace pattern was designed to be incorrectly detected as multiple faces. When printed on fabric, someone wearing it would create false positives for a particular face detection algorithm. Image credit: Adam Harvey.

or killed as a result. When AI audits produce evidence of flaws in AI and there is inaction to fix the system or identify underlying causes, these flaws can become a way for users or other unexpected parties to act on that information. This is why it is so crucial to take steps to address the problems revealed in an AI audit.

Audits Reveal Values

AI auditors face many situations where there is no one correct or objective output or score for an audit to judge. When an audit characterizes a system's actions, it can lead to debate about the system's purpose, its priorities, and whether it demonstrates values that reflect the interests and well-being of the people it affects. An audit report can make evident the values, stakes, or other previous decisions embedded in a system that were otherwise never public. And it can provide grounds for choosing between competing values or two alternative courses of action.

Audits Explore Alternative Futures

The United Network for Organ Sharing (UNOS) kidney transplant system introduced earlier was designed to rank people's relative needs for kidney transplants. The foundational problem in this case is that there are not enough transplant kidneys for the people who need them, and there is no obviously "right" way to allocate the kidneys available.

But UNOS's original system definitely used the wrong approach. The system's calculations included race as a factor, because of a decades-old false belief that people of different racial backgrounds had different levels of creatinine, a chemical compound that is used to measure kidney function.[22] This decision meant that Black people with kidney disease faced years of additional waiting on the transplant list. An audit of this system helped identify and concretize this disparity and revealed problems with other proposed allocation methods that had been considered as replacements.

As new information from the audit surfaced, the UNOS system had to change, and this required important decisions about what should be prioritized in kidney allocation.[23] Doctors, patients, and hospital administrators drew on the audit results to sketch out different possibilities and estimate the impact they would have on people's lives. The resulting algorithm demonstrated the power of creating dialogue between stakeholders in action. That the coalition was able to build a system that people could largely agree on, despite its life-or-death implications, is a remarkable success.

Audits Clarify Ambiguous Standards and Situations

Audits should be designed to answer specific questions. On mainstream social media platforms, content moderation systems are used to identify and weed out violent and hateful speech. But these systems are opaque. Most social media users cannot articulate what is prohibited on each platform. Even reading a platform's public rules will not make this clear. When a video gets taken down from TikTok, it is not necessarily clear why: These are matters of interpretation, and there is rarely a person inside the company who's ready and willing to explain how the rules are enforced. To make matters more complicated, this kind of moderation is increasingly handled automatically, by AI systems.

Audits of AI systems used by social media platforms can help reveal how content rules are being enforced (or not enforced). We know Reddit prohibits content that promotes hate based on identity or vulnerability.[24] Facebook (now Meta) has different rules for hate and misinformation.[25] In the United States, these prohibitions are corporate policy decisions—both of these prohibitions include legal speech. Content creators, parents, and other concerned parties want to know how those rules are applied.

Audits that scrutinize this kind of automated content moderation may ask specific questions, but they do not always have objective benchmarks and will likely traffic in highly subjective categories of content, such as what counts as hate. Audits in this area have revealed other hidden dimensions of content filtering and prioritization.

Some audits have unearthed decisions that platforms made deliberately but secretly, while other audits identified unintentional patterns of moderation that resulted from the complexity of an AI system—patterns of which the social media platform's users and maybe even the owners were completely unaware.

For example, audits can reveal that speech expressing support for one political position is systematically removed while the opposing side is permitted. Conversely, audits can prove this is not occurring, even when some politicians allege it is.[26] In such cases, the audit may serve an important descriptive role even if it does not lead to any repair or other change. Instead, the audits inform interested parties which platforms will allow different kinds of speech—and how abstract rules and prohibitions are being implemented in real situations.

Audits Reveal Choices and Motivate People to Reconsider Them

In other contexts, audits that reveal choices can be critically important in assessing the legitimacy of an AI tool. Consider risk scoring systems that are used to identify who should be detained in the criminal and immigration systems—they aim to quantify a person's relative "riskiness." However, investigations have revealed that such tools often embed policy choices or assumptions about values that are not based on evidence.

An audit of a risk tool called Risk Classification Assessment used by US Immigration and Customs Enforcement (ICE) to identify which immigrants to detain pending removal proceedings, for example, revealed that it permitted supervisor overrides for local policy reasons and committed all moderate-risk individuals to supervisor determinations. The tool thus obscured a pattern of unconstitutional detention. It was only when the audit was able to interrogate which characteristics mapped to "low" versus "moderate" risk that this was discovered and debated.[27]

In summary, by their nature AI tools will embed certain values and priorities, but these often remain hidden. An audit can make those values and priorities explicit and can provide evidence that can inform—but also structure—conversation and debate. In some cases, like those we turn to next, an audit may help drive changes in public policy.

Audits Influence Policy and Enforce Laws

Audits are playing an increasingly important role in public policy, sometimes prompting enforcement actions or revealing problems that inspire new policies. Governments have also mandated audits in sensitive areas where individual rights and welfare are at stake and subject to an AI system. Audits are a basic, necessary component for public accountability.

One definition of public policy is that it is the government's response to societal problems. The most impactful audits reveal harm in easily understandable terms and rhetorically powerful ways that have policy implications. As more audits have been carried out, they have collectively helped inform overall policy toward AI. In 2016, the Obama White House highlighted key AI audits in a call to address civil rights violations exacerbated by automated systems.[28] In 2021, Biden administration technology policy advisors engaged in a consultation process that would yield the White House *Blueprint for an AI Bill of Rights* in October 2022. This influential policy document, which drew on the expertise of AI audit researchers, would become a cornerstone of Biden administration AI policy, including the then-president's 2023 executive order on the Safe, Secure, and Trustworthy Development and Use of Artificial Intelligence.[29] While the executive order was rescinded by the Trump administration in 2025, the AI Bill of Rights continues to serve as a model for state, federal, and transnational regulation. These policy documents relied on the research on audits and recommended them as a key tool for evaluating AI systems. What's more, recent legislative proposals have sought to mandate more audits for any company using AI in critical decision-making.[30] For example, some policymakers require audits in laws that apply to individual industries, such as New York City's Local Law 144 for the jobs sector.[31]

Audits Support Enforcement

Audits will also continue to be used (just like old fashioned in-person housing audits) to enforce laws. To serve that function, audits need to provide specific information about a system's problems. This enables private parties to institute civil proceedings to vindicate their rights and also enables government regulators to investigate a system and begin enforcement measures.

In cases of fraud, audits can demonstrate when an AI system does not do what it claims to do. AI systems with a high error rate could violate constitutional due process rights where they affect people's liberty and employment. Audits that reveal bias based on protected characteristics such as race and sex help to prove violations of antidiscrimination laws in housing, employment, or credit. Audits that examine value or policy choices might reveal that the AI system is illegally considering factors about people that violate their rights—like someone's race, for instance, in decisions about who is arrested or incarcerated. These are just some examples of how audits are already used to enforce existing laws.

For audits to inform civil cases, they typically need to demonstrate what lawyers call standing—proof that the plaintiffs have suffered harm from the system or are likely to. A recent civil lawsuit against Facebook over its targeted advertising system cited an audit that found that Facebook ads for housing and jobs were skewed against women.[32]

The plaintiff claimed that, as indicated in the audit, Facebook's AI system harmed women as a protected class.

US law allows testers to have standing to bring claims to enforce civil rights laws like the Fair Housing Act, even if they were not actual applicants for housing who were discriminated against.[33] If someone participates in a collaborative audit by testing a platform's services, they could sue to enforce antidiscrimination laws.

Even when audits do not pinpoint a specific individual or class of people who were harmed, they can still influence government enforcement. Regulatory bodies may rely on an audit's outline of a problem to investigate a system further, conducting their own audit. Even if the audit does not reveal everything that regulators need to know, they are sometimes entitled to further inside information about the AI system after opening an investigation. This occurred when the US Department of Housing and Urban Development (HUD), drawing on an external audit, launched legal proceedings against Facebook for its ad delivery system. This tripwire audit gave HUD enough information to conduct its own investigation and file a complaint of discrimination that resulted in a settlement with Facebook.[34]

Audits also enable regulators to make claims about what a company knew about problems with its AI systems. When companies claim that well-documented problems were unanticipated, information from published audits can tell a different story. In HUD's complaint against Facebook alleging discrimination in ad delivery, the agency noted that the company was aware of the publicly available academic audit that prompted HUD's investigation.[35] Thus, the audit can serve as evidence of a company's lack of corrective action if the courts become involved.

Dueling Audits in the Courtroom

Audits that initiate court cases can lead to more detailed audits of a system's inner workings. During the discovery phase of a lawsuit, when evidence is gathered and witnesses are identified, courts may require companies to turn over further information about AI systems. This can include formulas and data sources that companies would otherwise keep confidential in the name of trade secrets or privacy. That information can then support audits by experts hired by either side in the litigation, whether to defend the company or to prove a system is violating the law.

Dueling courtroom audits were central in a 2012 case brought by people whose benefits were cut by the Idaho Medicaid program.[36] Idaho developed an AI system that claimed it could predict the needs of the patient population, but the historical data used was flawed and only a small portion of it was usable.[37] When Idaho cut benefits for people with developmental disabilities, the state argued that its benefits algorithm was a trade secret and could not be inspected. When courts ordered the details to be revealed to both teams of lawyers, each side produced a competing expert report assessing the AI system's performance. The court ultimately issued an order requiring the state to stop using the AI system and the parties settled the case, with the state agreeing to abandon that AI system and create a new one. The type of expert report produced in the Idaho case is common in civil discovery. As courts increasingly consider challenges to AI systems in a variety of contexts, the need for auditors who can perform the assessment and serve as expert witnesses will increase as well.

Even without litigation, government regulators can issue demands to gain internal data and information about a system and then audit it. The New York State Department of Financial Services did just that in 2021 when it was alerted by consumers that the Apple Card credit card seemed to be offering women lower credit limits or denying them accounts unfairly, possibly due to bias in an AI underwriting system.[38] The department launched an investigation that included thousands of pages of documents provided by Apple and the underwriting bank. The department then ran its own analysis on a dataset of 400,000 New York credit applicants which concluded that there was no disparate treatment or impact in the Apple Card's lending practices and also concluded that the bank's underlying statistical model did not consider prohibited characteristics of applicants or produce disparate impacts.[39] These audits can be powerful tools to reveal that there was no violation of law where one was suspected.

Audits as Opportunities to Organize the Public

In 2015 a congressional committee found that hundreds of millions of Americans were affected by data breaches and other privacy problems. An emerging market of data brokers—then a new kind of company—was buying, selling, and, unfortunately, leaking people's sensitive data in massive data breaches. Misuses of this data exposed large numbers of people to fraud, scams, and other manipulative practices that cost billions of dollars annually.[40] But then, in 2018, California passed a law requiring companies to disclose stored personal data. California was giving anyone a chance to find out how their data was used and request that it be deleted.

Auditors seeking to understand California's new law faced a significant challenge: With so many different companies holding private data, how could anyone possibly study them all? Enter Consumer Reports, a consumer protection organization founded in 1936, a time when product quality could mean survival or starvation for millions of Americans. Staff scientists at Consumer Reports regularly ask the organization's members to inform them about topics to investigate in order to produce product ratings and policy advocacy.[41]

When a team at Consumer Reports considered the question that California's new law prompted, it realized that the organization's six million members might help them find an answer. They already had connections with people across the country who were passionate about helping consumers, companies, and regulators by scientifically testing products. Why not apply this approach to data privacy?

For two months in 2020, Consumer Reports invited members to look up data brokers, check whether the companies offered any way to comply with California's new law, and contact companies to exercise their rights under the law. Volunteers checked hundreds of data brokers and found frequent cases where companies did not give consumers options to request their data or to remove it from the system. When people did make those requests, many companies never took action or never confirmed that action had been taken. Informed by its members, Consumer Reports published a report for regulators and companies looking for ways to improve their practices.

This volunteer audit by Consumer Reports gave everyone in the story direct value. Consumers had a chance to learn how their data was being used and manage it more directly, coached by the team at Consumer Reports. The California government got an early look at how the law was being used, which informed how it recruited and built the new California Privacy Protection Agency the law created. Companies also received valuable feedback on best practices for complying with the law, based on the experiences of over five hundred people who tested the process.[42]

In the years since this volunteer audit, Consumer Reports has built these kinds of audits into a regular part of its services for members. In 2023, it launched Permission Slip by CR, an app people can download to manage how data brokers handle their data. If people grant permission, the app also shares data with Consumer Reports that can be used for further advocacy and organizing. The free app has been so valuable that it motivates people to become paid members of Consumer Reports, growing the network of people supporting advocacy and improvements across AI.[43]

Audits Strengthen Advocacy

This is a model that also applies to AI audits. AI audits sometimes create enough outcry that social movements form around them. When the Gender Shades audit revealed racial and gender bias in facial analysis systems, it led to widespread complaints and media coverage.[44] Some companies responded by attempting to increase the accuracy of facial recognition for Black women. But many people, including Buolamwini herself, see increased accuracy as a problem rather than a solution. Black leaders in public policy and academia have argued that a fully functioning, highly effective facial recognition system could actually infringe on their privacy and perpetuate surveillance that could lead to further harm. In other words, it isn't clear that people wanted facial recognition software to work well. The conversation around this audit attracted volunteers and funders to a growing network of racial justice and privacy organizations that had not previously been as visible or powerful.

Whether audits are taken up by formal organizations or loose networks of advocates, they can crystallize debate and connect people. In this way, AI audits are like the earliest civil rights audits, which inspired partnerships that both drew from and strengthened existing community organizations and helped create new ones, leading to new not-for-profit auditing organizations in housing.[45] These communities can then also take their understanding of the issues into the policy domain. A completed audit, then, can give rise to new issue-focused consultancies, not-for-profit educational and advocacy organizations, or even new state agencies.

The Consequences of Resisting and Ignoring Audits

When someone discovers a misbehaving AI system and complains, the system's operator is often tempted to push back. That happened with Sarah Chase, the vice president of the National Eating Disorders Association (NEDA), during a pilot of its AI mental health chatbot. When told about bad advice the chatbot was giving to people with eating disorders, Chase initially responded on Instagram, "That is a flat-out lie." As public pressure mounted, she quickly changed her approach, asking people to "please send the screenshots—and if this is happening in the program, the screenshots will be essential to fixing it." NEDA's subsequent investigation confirmed the reports and led it to end its AI chatbot.[46] Chase's initial resistance to reported problems was not a flattering image for NEDA.

In 2019, when an audit revealed substantial racial bias in an AI healthcare system that gave care priority to healthier White patients over more seriously ill Black patients, Optum, the company that developed the system, characterized the audit findings as "misleading."[47] After reading the audit, the New York Department of Health and the state's Department of Financial Services demanded that the company either prove the AI system was not racially biased or stop using it.[48]

Organizations have many short-term reasons to deny problems. Leaders may want to avoid the legal risk of admitting a problem that exposes them to liability. They might worry that fixing an issue will be prohibitively expensive. Leaders may believe the potential harm to their reputation caused by mistakes would be worse than

When told about bad advice the chatbot was giving to people with eating disorders, the VP initially responded, “That is a flat-out lie.” As public pressure mounted, she changed her approach.

continuing to operate with a broken system. Or they might believe that the problem revealed by an audit is too insignificant to warrant action. But companies also risk reputational damage and legal liability when resisting or ignoring audits.

AI developers sometimes try to stop an audit while it is underway. But obstructing an audit tends to motivate your critics. For example, Mottahare Elsami, one of the authors of this book, led an audit of the AI system used by Yelp to filter and highlight reviews. Yelp sent a warning message to the team threatening to ban them from the platform if they did not stop contacting Yelp users. The researchers found a way to complete the study anyway. Simply making auditors' lives more difficult is unlikely to work. Circumspect auditors will certainly avoid making baseless accusations about the systems they investigate, but they are unlikely to abandon an investigation into serious wrongdoing or harm merely because the target doesn't like it.

Since audit researchers are unlikely to be deterred, we encourage companies to learn from NEDA's example by acknowledging the possibility of problems and committing to thorough investigations when auditors raise issues. Rather than responding with mistrust, companies can act in ways that build support for an accurate public understanding of AI's capabilities.

* * *

This chapter reviewed instances where audits led to embarrassing denials that were later retracted, or to elaborate patterns of dueling audits. Overall, we have surveyed the many kinds of impacts that follow an audit. Audits can help improve products, inform the public, trigger new regulations, and even reveal fundamental values embedded in the system. Audits lead to new partnerships and organizations and have given detractors creative ammunition that is used to attack AI systems (such as adding intentional errors). But what does it take to nourish a sustainable culture of audits? We have shown what is required for AI audits to provide real accountability. The next chapter describes what is needed to support a healthy AI ecosystem where AI audits are routine.

6

A HEALTHY AUDIT ECOSYSTEM

Think about the last glass of water you drank. If you live in a region with a safe water supply, you likely did not stop to think about whether it might make you ill. People who use this water system do not need to worry about whether the groundwater is contaminated or if the pipes are unsafe. Instead, drinking water is unremarkable, thanks to regular tests at each step in the system that prompt changes when needed, from the reservoir to the plumbing to the products that filter pollutants.

In a healthy AI ecosystem, encounters with AI should be as safe and reliable as drinking water from a clean tap. Yet, there is a great deal of public anxiety about AI tools available today, and for good reason: Many examples of problematic AI featured in this book fuel people's concerns about these technologies. AI auditors envision and work toward a world where AI is reliable, unremarkable, and safe.

As this book conveys, it has become clear that AI systems must be paired with audits and other mechanisms of AI accountability to earn public trust. Most Americans believe that AI will destroy more jobs than it creates and that it will interfere with democratic processes. Their concerns about new uses for AI far outweigh their excitement. And such concern is increasing.[1] In this chapter, we explore the ecosystem necessary for AI audits to be routine and unremarkable.

The Culture of Big Tech

In 2010, the Facebook corporation (now Meta) produced a series of posters. With bold red lettering on a cream background, each poster offered a single arresting phrase. One trumpeted the company's official motto from that period: "Move Fast and Break Things." Others read: "Done Is Better Than Perfect," "Orville Wright Did Not Have a Pilot's License," "Nothing Is Riskier than Not Taking Risks," and "What Could Go Wrong?" (figure 13).[2]

These motivational phrases adorned the walls of Facebook offices in the San Francisco Bay Area, in an attempt by the company to reinforce its corporate values.[3] It would be unimaginable to see such posters like these hanging in the halls of an aviation firm or at a nuclear reactor operator. But this nevertheless was and still is the normative posture

Figure 13 Facebook (now Meta) CEO Mark Zuckerberg stands in front of the company's then-motto at a technology conference in 2014. Photo credit: Mike Deerkoski (CC BY 2.0).

of many technology companies that enjoy soaring profits.[4] Today's AI push has increased criticism, but in many ways, the "What Could Go Wrong?" culture has remained.[5]

In the past, social media was largely considered a low-stakes industry whose services did not merit the same scrutiny as airplanes or nuclear reactors. But today it is clear that a misbehaving AI decision tool within a social network could facilitate a teen's suicide, accidentally force a queer person out of the closet, subjecting them to violence or even arrest, or become a vehicle for a foreign influence campaign. The risks of these systems are now frighteningly clear. How can the AI ecosystem responsibly earn public trust, even when its systems risk causing serious harm?

Sociologist Charles Perrow famously coined the term *normal accidents* to describe failures of technological systems in domains like aviation, civil engineering, and nuclear power.[6] For Perrow, these failures are normal—meaning, by his definition, inevitable—because the technologies have features that make accidents more likely. In these complicated systems, even if designers cannot predict what specific problem will arise, they must expect some issues to occur and plan accordingly.

Today's AI systems share some attributes with the technologies that concerned Perrow. They often have interactive complexity, with many components dependent on one another. Their problems can be difficult to foresee. They are very widely used. They operate quickly, and they have the potential for catastrophic failure, with human lives at stake. In these circumstances, problems can propagate rapidly, relatively invisibly, and in ways that are not under central control. In other words, some AI systems are prone to normal accidents.

The stakes could not be higher for AI. For more than a decade, the US Department of Defense has been known to use AI to find and kill targets. It is engaged in ongoing contracts to build new AI systems for military use with major companies including Google, Anthropic, and Palantir. A 2012 National Security Agency presentation showed how SKYNET, a drone strike system, analyzed cell phone and social media data from across Pakistan to identify potential extremists to target and kill. Investigative journalists alleged that SKYNET-targeted strikes resulted in up to four thousand deaths in Pakistan, many of whom were civilians.[7] After reviewing the SKYNET presentation, outside experts found the design of the system as it was depicted to be "scientifically unsound."[8] As we prepared to write this book, investigative journalists in Israel had begun publishing stories about Lavender, an AI system that the Israel Defense Forces was using to target members of Hamas in Gaza with airstrikes, in which civilian casualties were factored in as part of the system's "error rate."[9]

When Charles Perrow developed the idea of normal accidents to address the risk of nuclear reactors and other dangerous technological systems, he was pessimistic about our ability to respond to risky technologies, and he did not argue that we would be able to forestall all possible harms. In addition to preparing for failure, he predicted that the work required to handle some systems' problems would be so great as to outweigh the benefit of the system in the first place, and that some managers would instead choose to shut these systems down.

In technologies that feature normal accidents, Perrow also linked normalcy to another problem: an unnerving managerial culture in these industries that was not effectively guarding against risk. He argued that in corporate settings where accidents can easily become a routine expectation, organizations need ways to resist becoming complacent to the threats presented by potentially dangerous systems.

Initiatives to mitigate the inevitable problems—and sometimes fatal consequences—of AI systems should be planned and become routine. Perrow's analysis can inspire us to advocate for handling risky systems by integrating monitoring, including audits, so problems are made visible as soon as possible. Another tactic for handling risk is to be deliberate and slow—the opposite of "Move Fast and Break Things."

Audits Are Normal

A safer and more reliable AI ecosystem will build trust through routine, ongoing auditing. Continuous auditing is essential to adapt to evolving challenges and to ensure systems remain fair, transparent, and accountable.

A simple example of normal auditing might be that of legendary computing author Donald Knuth. Recognizing that errors in his books were unavoidable (Perrow might call them "normal"), he challenged readers to find mistakes, advertising in his prefaces a modest bounty of $2.56 to the first reader to find each one.[10] Knuth is a prolific writer and has signed error bounty checks totaling thousands of dollars over his career. In computer science, finding a problem in a Knuth book first is a badge of honor. Instead of cashing a check from Donald Knuth, recipients often save them as keepsakes.

Some software companies have also begun offering rewards for errors identified in their code, which they call *bug bounties* ("bug" in the sense of a software error).[11] Governments have even written bug bounties into regulations and policy documents.[12]

Effective bounty programs require a group of people capable of identifying errors, and they must be granted access to the artifact being studied. Some bug bounties also need access to specialized documentation and data before bounty hunters can find anything useful. Bounties also require legal infrastructure and companies willing to welcome and fund them. Without legal protections, those who come forward with errors would be right to fear being sued for defamation or hacking. This was once the case. Community and industry norms in software engineering were initially hostile toward those identifying errors, considering them hackers or adversaries, but today their presence is routinely welcomed. The same crucial shift is beginning to take place for AI audits.

Twitter's Bias Bounty

When Colin Madland, an education technology student at the University of Victoria in British Columbia, Canada, logged onto a Zoom video call with a professor at his university in the fall of 2020, he was astonished to see a headless body talking with him. His professor was suddenly invisible to his students because he was Black, and the video platform's AI did not recognize him. The erasure of Black faces from Zoom was hindering education at the height of the pandemic.

Looking for a solution online, Madland posted pictures to social media about the problem, showing photos of the professor with and without a visible head. His post spread widely on Twitter, but not for reasons he expected. At the time, Twitter used facial recognition AI to crop images and make faces more prominent. Adding insult to injury, when Madland posted about Zoom's AI, Twitter's image cropping AI cut out his professor too. Madland wondered if the AI systems at Twitter were prioritizing faces with lighter skin colors—a problem other people had also noticed. In response to his thread, other users posted pictures, shared examples, and debated the causes. When journalists noticed the conversation and published stories about it, even more people joined in.

At the time, Twitter had recently hired Rumman Chowdhury to lead a new team focused on AI ethics. Guided by Chowdhury's team, Twitter's leadership embraced public curiosity behind the growing flood of complaints and tests. As a first step, Chowdhury's team conducted an internal audit of the site's image cropping AI. Next, the company commissioned a bias bounty challenge. Making the AI code and data publicly available, the challenge offered up to $3,500 to anyone who could produce the best audit of its system.

The audit and bias bounties earned Twitter widespread praise. Speaking later about the process, company executives described the work of cutting through legal issues and red tape at record speed to make resources available to bounty hunters.[13] Bias bounty hunters also discovered a number of additional problems with the system. For example, the winning entry by then-student Bogdan Kulynych demonstrated that the AI system preferred to highlight pictures of younger, thinner women with lighter skin (see figure 14).[14] Elon Musk purchased Twitter the year after this bounty and renamed it X; since then, X has not hosted any more bias bounties.

Abandoning the AI

Despite Twitter's efforts, the complexity and pervasiveness of the issues with the cropping system proved too costly to resolve. Concluding that the currently available technical solutions were insufficient to address the biases fully, Twitter chose to withdraw its automated image cropping tool from use, a choice we refer to in chapter 5 as

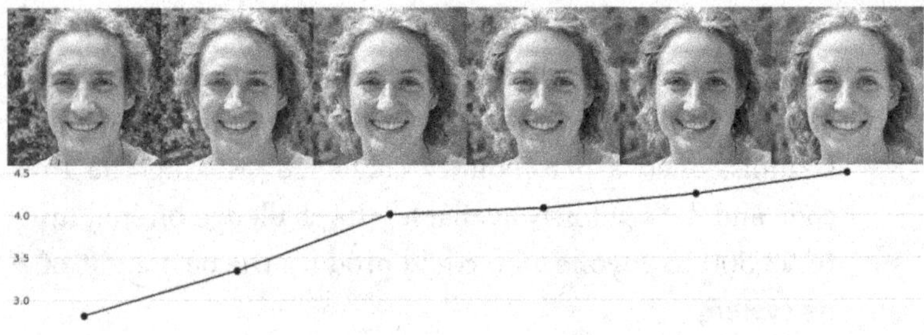

Figure 14 In this illustration from a Twitter audit, a photograph was artificially manipulated to make it more stereotypically feminine (from left to right). The more stereotypically feminine it became, the greater the chance that the Twitter image cropping algorithm selected the photo. Similar results were achieved by making faces more slender and lighter-skinned. Image credit: Bogdan Kulynych.

abandonment. Instead of AI, the company announced that it would allow users to choose how to crop their images, stating that people—not AI—were better equipped to make these decisions.[15]

Following Perrow's logic, we agree that AI audits, while usually framed as a compliance or accountability tool, can also help determine when an AI system should be shut down. What's known as abandonment can be framed as a positive step—by abandoning its image cropping AI, Twitter gave users the power to make their own decisions.

Red Teaming

Another practice of collective system investigation is known as *red teaming*. Developed in the context of computer security, external experts come together and form "red teams"—groups whose goal is to break into a system to identify its vulnerabilities. Drawing inspiration from those efforts, some have begun wrestling with AI systems' embedded harms (rather than security flaws) using the same approach. For instance, following Twitter's bias bounty, Rumman Chowdhury, the company's team lead for AI ethics, hosted a red teaming challenge at the 2023 DEF CON computer security conference to find such issues with a generative AI system like the one audited for hiring biases described in chapter 3.

Red teaming is not AI auditing (red teams may not have time for a complete, thorough, or systematic audit), but it can be used in tandem with, and sometimes closely resemble, the auditing approaches reviewed in this book. Routine red teaming is a complementary tactic for AI accountability and safety that should be pursued in parallel with routine audits. For example, the result from red teaming a system can be used to scope and determine the questions to be asked in an AI audit that further investigates.

Human Rights Impact Assessments

Multinational companies of many kinds—including mining, extractives, fashion, and technology—regularly undergo what are known as human rights impact assessments. These evaluations of a company's business activities and their effects are measured against a set of benchmarks drawing on international human rights doctrine and agreements including the UN Guiding Principles on Business and Human Rights, which covers everything from labor to data privacy rights. These principles are increasingly being used to inform AI governance and policymaking and can be useful for AI auditors seeking insights and guidance from accountability mechanisms used in other industries.[16]

In 2018, a UN fact-finding mission highlighted links between calls for violence on Facebook and ethnic cleansing operations by the Myanmar military in which an estimated 25,000 people of the Rohingya ethnic minority were killed. In response, Facebook commissioned BSR, an independent consultancy, to conduct a human rights impact assessment of its business practices in the country.[17] The resulting audit, for which BSR conducted interviews with sixty civil society, government, and private sector stakeholders in Myanmar, found that Facebook had become "a useful platform for those seeking to incite violence and cause offline harm" and that Facebook had not adequately enforced its community standards in Myanmar. Facebook published the assessment in full, signaling an effort to assess and understand its failures.[18]

These assessments are not audits, but they form part of the same broader ecosystem of accountability measures that third parties can carry out on their own, or that companies can commission in an effort to make themselves more accountable to the public interest.

Standardization and Sustaining Innovation

Many companies have shifted from seeing the bounty programs reviewed above as antagonistic and hostile to welcoming and facilitating them. But not all audits are received with open arms.

Auditors, Not Hackers

Twitter's audit team and its bias bounty provided an interesting blend of internal and external auditors working to solve a problem with an AI system. When outside auditors are involved, one of the requirements for normal auditing is clearing away legal obstacles to probing an AI system. The team at Twitter handled this problem with the bias bounty's legal agreements. But what happens when the outside auditors are not operating within this kind of framework, as is the case for investigative journalists investigating a system from the outside?

Antiquated computer crime laws that prohibit hacking closed computer systems have sometimes been applied to common methods used in external audits of open websites and platforms, even though such laws were never intended to stop people from doing things like looking at publicly available web pages. The most notorious of these laws is the federal Computer Fraud and Abuse Act (CFAA), which for years was understood by courts and the US Department of Justice as criminalizing any violation of a platform's terms of service. This was a significant problem for many independent auditors, who often study platforms whose terms of service expressly prohibit research.

Platforms' terms of service are written unilaterally by the system's developers and lawyers to serve their interests. They often prohibit common auditing techniques like those described in chapter 3, such as scraping public data or using sock puppet accounts with fictional user information.[19]

Faced with the threat of potential criminal liability, a group of academic researchers, including Karrie Karahalios and Christian Sandvig, represented by ACLU lawyers including Esha Bhandari (all authors of this book), filed a lawsuit arguing that the CFAA should not apply to their research audits of websites. They argued they had a First Amendment right to engage in such work and to publicly report their findings. That lawsuit, *Sandvig v. Barr*, was the first to argue that computer crime laws intended to outlaw "hacking" could not be applied to independent audits by researchers and journalists seeking to inform the public.[20]

The case was resolved in favor of the researchers when a court held that the CFAA did not criminalize violations of terms of service alone.[21] The US Supreme Court later provided greater clarity for auditors when it ruled that the CFAA did not apply to violations of written terms governing access to a system. In that ruling, the court specifically referenced *Sandvig v. Barr*, noting the importance of AI audits, including many of the audits described in this book. (The court referred to AI audits as "online civil rights testing."[22])

AI auditors have over eighty years of offline social science audits to draw from when considering the ethics of AI auditing. Cases like these show progress in establishing what legal scholars have called the "right to test" AI systems, and what we would call a right to audit.[23] Legal obstacles remain, but courts, regulators, the public, and companies are increasingly recognizing the important work of independent auditors and the need for robust legal protections for audits.

The Costs of an Audit

Audits are a testing process. Everyone's expectations for testing—and the costs it incurs—depend upon who the tester is. In a product testing process run by a manufacturer on its own product, that owner has complete control over the product it owns and operates. Any resources used in product testing are the owner's to spend. Outside auditors, by contrast, are effectively using someone else's resources, because they run their tests on someone else's system.

Audit studies are normal across all products and topics of research, from medicine to finance.[24] The consensus is that the stress placed on the audited system is appropriate when the question addressed in the audit is important (for society or for a group that may be harmed). The importance of the goal justifies the resources spent, particularly when there is no other way to evaluate an AI system.

Some firms may bristle at the reputational damage they face when an audit casts them negatively in the public eye uninvited and without warning. But housing auditors established decades ago that it is inappropriate to warn someone violating the law that you are going to try to catch them in the act. Indeed, seeking advance permission from racist landlords before a housing audit would ensure that no discrimination would ever be found.

Audit targets also may also argue that when auditors add false data to their system (such as with sock puppet audits), this effectively pollutes the system's datasets. However, skilled auditors have a variety of technical means to minimize the amount of test data that they add, or to remove their test data after the audit's completion. And many audits that insert data are undetectable. Moreover, any impact on the target system must be understood in the context of the amount of "pollution" that is typical in that system to begin with. When a system is open to the public, spammers, would-be hackers, bots, and the system's own users already add a great deal of material to it that is false or considered undesirable by the system's operator. In these systems, a well-designed audit inserts a small fraction of the data that spammers routinely add. The act of inserting data should be judged by the audit's impact on the system's usual operation.

Some review committees in academic institutions and nonprofits have guidelines that prohibit investigations of

Housing auditors established decades ago that it is inappropriate to warn someone violating the law that you are going to try to catch them in the act.

harm from *causing harm themselves*. Again, review committees that have approved social science and civil rights auditors for decades have found that this restriction does not apply to audits testing for serious problems. The injuries befalling a company whose AI is found to be dangerous are the just deserts of the company's actions, not the audit results that uncovered the problem.

Typical outside audits do not aim to protect the target's reputation or livelihood. The duty of the auditor is to society and the people affected by the system under scrutiny. Corporate and governmental ethical review boards that are informed about the history of auditing understand this. A robust, well-designed audit of a system should proceed so long as any costs caused by the audit are outweighed by its benefits to society.

Professionalization

Auditors used to be at risk of facing legal consequences when their work was unwelcome, but as the field grows, this dynamic is flipping. Instead, the AI companies themselves are beginning to become liable for the consequences of their products. As a result, the demand for AI audits is growing and the work is becoming increasingly professionalized. There is already a healthy push and pull between the value of reliable standards, certificates, and best practices and the need for innovative approaches to auditing black box systems.

The injuries befalling a company whose AI is found to be dangerous are the just deserts of the company's actions, not the audit results that uncovered the problem.

While the specifics depend on the jurisdiction, AI system developers will likely be subject to legal liability wherever they manufacture or sell their products.[25] US law is still evolving with respect to AI, but courts have already begun hearing cases about harms caused by AI (e.g., *Nilsson v. General Motors LLC*[26]). Faced with potential legal liability, AI developers may seek to rely on standardized audit practices and third-party certifications to demonstrate that reasonable care was exercised in system design. In cases where harm leads to litigation, courts will likely find such audits relevant when deciding whether the system's developer was negligent and should be held responsible for those harms.

The endgame for these efforts is unclear. The law is still developing, and experts selling their labor as auditors are still emerging. As discussed in chapter 4, the risk of consulting firms selling glorified checklists as audits is a concern we keep in focus. We hope that through this book and our other research, we can help guide auditing standardization toward being formalized in more meaningful, rigorous, and expansive ways.

Democratizing Auditing

Before Twitter stepped in to welcome its users to participate in its bias bounty, users had already begun independently investigating the system. These users publicized the issues they saw, hypothesized about causes, collected evidence, amplified important points made by other users, and eventually provoked an escalation to the offices of Twitter's executives.[27] Several important aspects of the platform unintentionally supported them in this behavior, aspects that could help democratize such user-led audits on other platforms too.

Collective Auditing Infrastructures

Before Twitter was purchased by Elon Musk in 2022, the platform's infrastructure enabled users to test the system and share their results with others. In other words, users could start an organic audit and work together to build collective knowledge because the platform itself provided a place for users to meet each other and discuss issues.

X no longer has these affordances. But good platform design of AI systems and external tools can help make user-driven audits a reality on other platforms. In 2021, the US National Science Foundation funded the National Internet Observatory initiative to collect data on the online behaviors of a large, representative sample of US residents and provide access to this data for AI auditors globally.[28] This observatory will provide auditors with validated data sources from users who have volunteered to share their behaviors, so the auditors do not need to start every new audit from scratch.

Today, civil society organizations, governments, and academics are building and operating infrastructure that supports and sustains people's collective auditing. It is ever more common for people to find each other online and share examples of strange or unlikely things that AI systems do, especially when they seem inaccurate or unjust. In the fall of 2023, WhatsApp users trying to send each other messages about Palestine discovered that when they prompted the system's visual AI tool to create a sticker representing the word "Palestinian," the tool repeatedly spit out images of young men and boys holding machine guns. Users coordinated to test the system and began posting screenshots of the stickers on social media. Within days, *The Guardian* had reported on the issue, and WhatsApp rushed to fix the problem.[29]

We are heartened by the level of knowledge and collaborative impulse that online communities exhibit in cases like these. Motahhare Eslami, an author of this book, leads an online platform called WeAudit that aims to facilitate action and aggregate voluntary effort among AI platform users to conduct collective audits. WeAudit targets problems of bias and fairness, particularly helping users scaffold their auditing processes to create effective and credible audits.

Audits Around the World

While this book is rooted in the context of the United States, we are keenly aware that the kinds of AI systems we discuss—and thousands of others—shape people's lives worldwide. Several of the audits we describe had ramifications for Google and Meta, both global companies whose products have been proven to have adverse impacts on users worldwide. There are also thriving tech markets around the world where new technologies are being built to perform these and countless other functions. And institutions like the World Bank are incentivizing national governments to institute automated systems to "streamline" and reduce the costs of government administration and social services. Qualitative audits by Human Rights Watch have already shown that inaccurate outputs from these systems are depriving people of much-needed benefits.[30]

As more governments adopt AI tools, they will need robust audits by independent parties, and to heed those audits' results. Amsterdam's recent attempt to create a welfare fraud detection algorithm illustrates both the promise and the peril of such efforts. Despite following established responsible AI practices—conducting bias tests, consulting stakeholders, and implementing technical safeguards—an investigation by Algorithm Audit, a Dutch nonprofit organization, concluded that the city's "Smart Check" system still exhibited bias. Ultimately officials abandoned the system after concluding that they could not definitively say the system was free from discrimination.[31] The Dutch case highlights a fundamental tension: holding AI systems to standards of fairness is necessary and may be difficult to achieve. Although much work lies ahead, we are heartened by the efforts of policymakers worldwide—from Brazil to Indonesia to the European Union—who are already building auditing requirements into policy regimes focused on AI.

What Kind of World Do We Want to Live In?

Computers, AI, and the internet can seem like mirrors of reality—tools that are reflections of society, and include all the same problems, from racism to price discrimination to fraud. But the constant development and deployment of untested new AI technologies that shape what we see and how we live can go beyond reflecting what exists.

In chapter 2, we described how Google's Gemini system generated historically inaccurate images of Black Nazi soldiers. In his attempt to explain what had happened, Google's senior vice president Prabhakar Raghavan described how developers had "tuned" the system to avoid biases that had stymied previous models.[32] But, he wrote, "our tuning to ensure that Gemini showed a range of people failed to account for cases that should clearly not show a range."

Google's Gemini exposed thorny questions about depictions of people—it wasn't easy to build a tool that could serve users who were trying to produce historical images, others who were creating speculative art, or still others who wanted to see representations of the present day.

Google chose a standard that failed to account for the wide range of situations in which people might use this tool.

This reveals a frequent, implicit disagreement about the goals for AI audits and these tools writ large. Do we want systems that attempt to mirror reality, showing us demographically accurate representations of people in different places or professions? Or do we want systems that offer a diverse and balanced portrayal of any chosen population? And how should we instruct these systems to represent history? Fact-checking this particular error in Google's output is easy, but identifying the right path forward—and how to build tools that allow us to pursue that path—is a far more complex challenge.

And this is just the most recent example. Google has struggled to determine how to handle material that might have historical or evidentiary value in the past. Video and image classifying AI used by YouTube to filter out videos of gratuitous violence have long been a subject of concern for people working to document war crimes in places like Syria, Yemen, and Sudan. Hundreds of thousands of such videos have been permanently taken down by YouTube's AI system, removals that international humanitarian law experts say obstruct efforts to prosecute the perpetrators of war crimes in all three places. These removals also deprive the public of documentation from critical periods in history. Without intervention, platforms like YouTube have the power to govern what we know and what we can discover about our past.[33]

Audits allow us to make informed choices about the kind of world in which we want to live, and they move important questions about our lives and collective values into public dialogue. Joy Buolamwini and Timnit Gebru's Gender Shades study proved that large-scale facial analysis systems failed to accurately label Black faces. Some tech companies quickly responded by gathering photos of Black people to build new datasets that could train their algorithms to more accurately label everyone, regardless of skin color. But not everyone wants to be recognized by facial recognition technologies. Do we want to live in a world where AI systems can identify each of us with razor-sharp precision? Or do we want to live with privacy, in a world where AI systems operate only with our consent, and with respect for our humanity?

Writing in exile from Turkey during the Second World War, the Jewish German philosopher Karl Popper asked what role information should play in free societies. In his book, *The Open Society and Its Enemies*, Popper argued that people in democracies need public information to work together for the common good. As AI systems increasingly shape whole swaths of our lives and livelihoods, audits are an ever more critical source of that information. The next step is up to all of us.

ACKNOWLEDGMENTS

The Center for Ethics, Society, and Computing (ESC) at the University of Michigan conceived and convened this project. The first draft of this manuscript was produced at Marquand House. Former resident Allan Marquand held the first professorship of art and archaeology at Princeton University and built a logic machine in 1887 that was a precursor to the computer. These facilities at the Institute for Advanced Study (IAS) in Princeton, New Jersey made this project possible to imagine and complete. Additional funding and support came from the IAS Science, Technology, and Social Values Lab, the University of Notre Dame–IBM Technology Ethics Lab, and the John D. and Catherine T. MacArthur Foundation. The words and ideas in this book are solely those of the authors. Producing a book with this many authors is no easy feat, and we are immensely grateful to Alysa Khouri and Faith Bosworth for helping us to tackle that challenge, and to Gita Manaktala for supporting the book.

GLOSSARY

AI
The capacity of computers to exhibit or simulate intelligent behavior in the broadest sense, or the systems that produce such behavior. Also called an AI system.

AI audit
A rigorous and systematic evaluation of the core claims, expectations, or standards pertaining to an *AI* system. Also called an *algorithm* audit.

algorithm
A procedure used by a computer to solve a problem, or, more generally, a computer system as a whole that is employing such a procedure.

auditor
The person or group conducting an audit.

audit study
As distinguished from a financial or tax audit, in social science, changing the inputs to a situation (e.g., by employing different *testers*) in order to evaluate any resulting changes in the outputs. For computing, see *AI audit*.

audit-washing
Using an ineffective *AI audit* to provide a system with the appearance of accountability and respectability, without the substance. Used with negative connotations.

bias
To influence or distort a system away from an expected, widely agreed-upon, neutral, ideal, natural, or objective behavior. Also the amount of such influence.

bias bounty
A reward offered, often to the public, for the discovery of a previously unknown bias. Cf. *bug bounty*.

black box
Any component of a system whose internal mechanism may not be readily known. Cf. *glass box*.

bug bounty
A reward offered, often to the public, for identifying errors in software. Usually intended to detect security vulnerabilities. Cf. *bias bounty*.

correspondence audit
A social science *audit study* that was once accomplished by sending letters in the mail (historically, cover letters and resumes). More broadly, shorthand for any audit study.

Crandall's Complaint
No reasonable person should expect a system to be unbiased. Named after the former CEO of American Airlines who defended the SABRE airline reservation system by admitting it was intentionally biased.

four-fifths rule
In employment law, if the selection rate for one group is less than 4/5 (or 80 percent) of the rate for the largest group, this can be taken as evidence of a substantial difference, or of discrimination. This rule was developed as a simple, practical test, but it is also arbitrary.

generative AI
A *machine learning* system focusing on pattern creation that is designed to produce output that has never existed before, including text, images, audio, and/or manufactured data.

glass box
Any component of a system whose internal mechanism is intentionally revealed. Also called a white box. Cf. *black box*.

ground truth
In *machine learning* and *AI*, data chosen to represent the standard that the system is tested against, or the data defined as true for the purpose of developing the system. These data are unlikely to be actually true, but they may be the best data available.

harm
Damage, injury, or loss suffered by a person, group, system, company, institution, or object. Also, to do harm.

impact assessment
A formal procedure that assesses the social, environmental, and/or economic consequences of some action. May be required by law or as an industry standard. Also called an impact report.

large language model (LLM)
AI that uses *machine learning* for language generation (e.g., OpenAI's GPT-4). Used in contemporary AI systems that interact with people using language.

machine learning
AI that infers patterns in a set of data to develop, change, or adapt without following explicit instructions.

muffin–Chihuahua problem
A situation where an *AI*'s performance may differ substantially from a human's, sometimes in a way that is not obvious. For example, a child sees the difference between a muffin and a Chihuahua, but an image recognition *algorithm* may not.

normal accidents
The sentiment that failures should be expected and may be unavoidable. Broadly, the position that planning for and mitigating failures is more useful than trying to design systems that will not fail. Named after sociologist Charles Perrow's book of the same name.

redlining
Systematic, often intentional, racial discrimination. Refers to the historical practice in banking of literally drawing a red line on a map to indicate Black neighborhoods where loans, mortgages, or insurance would not be offered.

red team
From computer security, a group pretending to be an adversary, usually hostile hackers, as part of an exercise.

scraping
Using a computer program to copy a large amount of information, often information that is publicly available. As web scraping, downloading a large number of web pages automatically.

sexy construction worker problem
The situation where members of a group are underrepresented and they are also misrepresented. Named for the finding that AI-produced images of female construction workers are sexualized and unrealistic, while images of male construction workers are not.

sock puppet
A computer script or program masquerading as a person. In an *AI audit*, an automated *tester*.

source code
The most human-readable form of a computer program.

target
The person, system, or company being investigated by an *audit study*. Also called the auditee.

tester
Historically, a human investigator who pretends to want an apartment or a job in order to conduct an *audit study* of a landlord or employer for *bias*. Later audits used letters (*correspondence studies*) or *sock puppets* instead of humans.

training data
Inputs initially given to a *machine learning* system from which it can infer patterns, sometimes described as a set of examples. Distinct from data used for validation, testing, or any other purpose.

NOTES

Chapter 1

1. "Kidney Disease Statistics for the United States," National Institute of Diabetes and Digestive and Kidney Diseases, last reviewed September 2024, https://www.niddk.nih.gov/health-information/health-statistics/kidney-disease.

2. "The Kidney Project," University of California San Francisco, accessed August 14, 2025, https://pharm.ucsf.edu/kidney/need/statistics.

3. "Need Continues to Grow," Organ Procurement & Transplantation Network, accessed August 14, 2025, https://optn.transplant.hrsa.gov/need-continues-to-grow.

4. R. A. Wolfe, K. P. McCullough, D. E. Schaubel, J. D. Kalbfleisch, S. Murray, M. D. Stegall, et al., "Calculating Life Years from Transplant (LYFT): Methods for Kidney and Kidney-Pancreas Candidates," *American Journal of Transplantation* 8, no. 4, pt. 2 (2008): 997–1011, https://doi.org/10.1111/j.1600-6143.2008.02177.x.

5. Sachin Waikar, "Evolution of an Algorithm: Lessons from the Kidney Allocation System," Stanford University Human-Centered Artificial Intelligence, April 27, 2020, https://hai.stanford.edu/news/evolution-algorithm-lessons-kidney-allocation-system.

6. Peter G. Stock and Ken Andreoni, *OPTN/UNOS Kidney Transplantation Committee Report to the Board of Directors* (Organ Procurement and Transplantation Network, 2009), https://web.archive.org/web/20100527112723/http://optn.transplant.hrsa.gov/CommitteeReports/board_main_KidneyTransplantationCommittee_7_24_2009_10_41.pdf; David G. Robinson, *Voices in the Code: A Story About People, Their Values, and the Algorithm They Made* (Russell Sage Foundation, 2022).

7. Cindy Alexander and Yoon-Ho Lee, "The Economics of Regulatory Reform: Termination of Airline Computer Reservation System Rules," *Yale Journal on Regulation* 21, no. 2 (2004): 379, https://openyls.law.yale.edu/handle/20.500.13051/8045.

8. United States v. American Airlines Inc. and Robert L. Crandall, 570 F. Supp. 654 (N.D. Tex. 1983).

9. Thomas Petzinger Jr., *Hard Landing: The Epic Contest for Power and Profits That Plunged the Airlines into Chaos* (Crown Currency, 1996).

10. Christian Sandvig, Kevin Hamilton, Karrie Karahalios, and Cedric Langbort, "Auditing Algorithms: Research Methods for Detecting Discrimination on Internet Platforms," paper presented at Data and Discrimination: Converting Critical Concerns into Productive Inquiry, a preconference at the 64th Annual Meeting of the International Communication Association, Seattle, May 22, 2014, https://websites.umich.edu/~csandvig/research/Auditing%20Algorithms%20--%20Sandvig%20--%20ICA%202014%20Data%20and%20Discrimination%20Preconference.pdf; Brigitte Alfter, Ralph Müller-Eiselt, and Matthias Spielkamp, "Automating Society: Taking Stock of Automated Decision-Making in the EU," *Algorithm Watch*, January 2019, https://algorithmwatch.org/en/automating-society-2019/.

11. For an overview, see Department of Transportation, "Computer Reservation System (CRS) Regulations," *Federal Register* 69, no. 4 (2004): 976–1033, https://www.ecfr.gov/current/title-14/chapter-II/subchapter-A/part-255.

12. Alex Barinka, "Meta's Instagram Users Reach 2 Billion, Closing In on Facebook," *Bloomberg*, October 26, 2022, https://www.bloomberg.com/news/articles/2022-10-26/meta-s-instagram-users-reach-2-billion-closing-in-on-facebook.

13. Julia Angwin and Terry Parris Jr., "Facebook Lets Advertisers Exclude Users by Race," *ProPublica*, October 28, 2016, https://www.propublica.org/article/facebook-lets-advertisers-exclude-users-by-race.

14. US Department of Justice, "Justice Department Secures Groundbreaking Settlement Agreement with Meta Platforms, Formerly Known as Facebook, to Resolve Allegations of Discriminatory Advertising," press release 22-650, June 21, 2022, updated February 6, 2025, https://www.justice.gov/opa/pr/justice-department-secures-groundbreaking-settlement-agreement-meta-platforms-formerly-known.

Chapter 2

1. Danaë Metaxa, Joon Sung Park, Ronald E. Robertson, Karrie Karahalios, Christo Wilson, Jeff Hancock, et al., "Auditing Algorithms: Understanding Algorithmic Systems from the Outside In," *Foundations and Trends in Human–Computer Interaction* 14, no. 4 (2021): 272–344, https://doi.org/10.1561/1100000083; Christian Sandvig, Kevin Hamilton, Karrie Karahalios, and Cedric Langbort, "Auditing Algorithms: Research Methods for Detecting Discrimination on Internet Platforms," paper presented at Data and Discrimination: Converting Critical Concerns into Productive Inquiry, a preconference at the 64th Annual Meeting of the International Communication Association, Seattle, May 22, 2014, https://websites.umich.edu/~csandvig/research

/Auditing%20Algorithms%20--%20Sandvig%20--%20ICA%202014%20Data %20and%20Discrimination%20Preconference.pdf.

2. Joshua Asplund, Motahhare Eslami, Hari Sundaram, Christian Sandvig, and Karrie Karahalios, "Auditing Race and Gender Discrimination in Online Housing Markets," *Proceedings of the International AAAI Conference on Web and Social Media* 14, no. 1 (2020): 24–35, https://doi.org/10.1609/icwsm.v14i1 .7276.

3. Sandvig et al., "Auditing Algorithms"; Devah Pager and Bruce Western, "Identifying Discrimination at Work: The Use of Field Experiments," *Journal of Social Issues* 68, no. 2 (2012): 221–223, https://doi.org/10.1111/j.1540-4560 .2012.01746.x.

4. S. Michael Gaddis, ed., *Audit Studies: Behind the Scenes with Theory, Method, and Nuance* (Springer, 2018); Metaxa et al., "Auditing Algorithms."

5. Marianne Bertrand and Sendhil Mullainathan, "Are Emily and Greg More Employable Than Lakisha and Jamal? A Field Experiment on Labor Market Discrimination," *American Economic Review* 94, no. 4 (2004): 991–1013, https://doi.org/10.1257/0002828042002561.

6. Sandvig et al., "Auditing Algorithms."

7. Jeff Larson, Surya Mattu, Lauren Kirchner, and Julia Angwin, "How We Analyzed the COMPAS Recidivism Algorithm," *ProPublica*, May 23, 2016, https://www.propublica.org/article/how-we-analyzed-the-compas-recidivism-algorithm.

8. Julia Angwin, Jeff Larson, Surya Mattu, and Lauren Kirchner, "Machine Bias," *ProPublica*, May 23, 2016, https://www.propublica.org/article/machine -bias-risk-assessments-in-criminal-sentencing.

9. Ben Green and Yiling Chen, "Disparate Interactions: An Algorithm-in-the-Loop Analysis of Fairness in Risk Assessments," *Proceedings of the Conference on Fairness, Accountability, and Transparency* (2019): 90–99, https://doi.org /10.1145/3287560.3287563.

10. Tiffany Wenting Li, Silas Hsu, Max Fowler, Zhilin Zhang, Craig Zilles, and Karrie Karahalios, "Am I Wrong, or Is the Autograder Wrong? Effects of AI Grading Mistakes on Learning," *Proceedings of the 2023 ACM Conference on International Computing Education Research* 1 (2023): 159–176, https://doi.org /10.1145/3568813.3600124.

11. Jeremiah Scanlan, "Auditing Predictive Policing," *Brigham Young University Prelaw Review* 33 (2019), https://scholarsarchive.byu.edu/byuplr/vol33 /iss1/4/; Johana Bhuiyan, "LAPD Ended Predictive Policing Programs Amid Public Outcry. A New Effort Shares Many of Their Flaws," *The Guardian*, November 8, 2021, https://www.theguardian.com/us-news/2021/nov/07/lapd -predictive-policing-surveillance-reform; Hamid Khan and Pete White, "Police

Surveillance Can't Be Reformed. It Must Be Abolished," *Vice*, March 10, 2021, https://www.vice.com/en/article/xgzj7n/police-surveillance-cant-be-reformed-it-must-be-abolished.

12. US Department of Justice, "Justice Department Secures Groundbreaking Settlement Agreement with Meta Platforms, Formerly Known as Facebook, to Resolve Allegations of Discriminatory Advertising," press release 22-650, June 21, 2022, updated February 6, 2025, https://www.justice.gov/opa/pr/justice-department-secures-groundbreaking-settlement-agreement-meta-platforms-formerly-known.

13. Matthew Kay, Cynthia Matuszek, and Sean A. Munson, "Unequal Representation and Gender Stereotypes in Image Search Results for Occupations," *Proceedings of the 33rd Annual ACM Conference on Human Factors in Computing Systems* (2015): 3819–3828, https://doi.org/10.1145/2702123.2702520.

14. Danaé Metaxa, Michelle A. Gan, Su Goh, Jeff Hancock, and James A. Landay, "An Image of Society: Gender and Racial Representation and Impact in Image Search Results for Occupations," *Proceedings of the ACM on Human–Computer Interaction* 5, no. CSCW1 (2021): 1–23, https://doi.org/10.1145/3449100.

15. Mauricio Chandler, "Streaming Stress: Why Spotify Is Receiving Artist Backlash," *Business Review at Berkeley*, August 24, 2022, https://businessreview.studentorg.berkeley.edu/streaming-stress-why-spotify-is-receiving-artist-backlash/.

16. Nick Seaver, *Computing Taste: Algorithms and the Makers of Music Recommendation* (University of Chicago Press, 2022).

17. Adi Robertson, "Google Apologizes for 'Missing the Mark' After Gemini Generated Racially Diverse Nazis," *The Verge*, February 21, 2024, https://www.theverge.com/2024/2/21/24079371/google-ai-gemini-generative-inaccurate-historical.

18. Metaxa et al., "An Image of Society."

19. Metaxa et al., "Auditing Algorithms."

20. William Dieterich, Christina Mendoza, and Tim Brennan, *COMPAS Risk Scales: Demonstrating Accuracy Equity and Predictive Parity* (Northpointe, 2016), https://go.volarisgroup.com/rs/430-MBX-989/images/ProPublica_Commentary_Final_070616.pdf.

21. Christo Wilson, Avijit Ghosh, Shan Jiang, Alan Mislove, Lewis Baker, Janelle Szary, et al., "Building and Auditing Fair Algorithms: A Case Study in Candidate Screening," *Proceedings of the 2021 ACM Conference on Fairness, Accountability, and Transparency* (2021): 666–677, https://doi.org/10.1145/3442188.3445928.

22. Meg Young, Michael Katell, and P. M. Krafft, "Confronting Power and Corporate Capture at the FAccT Conference," *Proceedings of the 2022 ACM Conference on Fairness, Accountability, and Transparency* (2022): 1375–1386, https://doi.org/10.1145/3531146.3533194.
23. Kyra Yee, Uthaipon Tantipongpipat, and Shubhanshu Mishra, "Image Cropping on Twitter: Fairness Metrics, Their Limitations, and the Importance of Representation, Design, and Agency," *Proceedings of the ACM on Human–Computer Interaction* 5, no. CSCW2 (2021): 1–24, https://doi.org/10.1145/3479594.
24. Jeffrey Dastin, "Insight—Amazon Scraps Secret AI Recruiting Tool that Showed Bias Against Women," *Reuters*, October 10, 2018, https://www.reuters.com/article/idUSKCN1MK0AG/.

Chapter 3
1. Melissa Heikkilä, "Joy Buolamwini: 'We're Giving AI Companies a Free Pass,'" *MIT Technology Review*, October 29, 2023, https://www.technologyreview.com/2023/10/29/1082632/joy-buolamwini-were-giving-ai-companies-a-free-pass/.
2. Joy Buolamwini and Timnit Gebru, "Gender Shades: Intersectional Accuracy Disparities in Commercial Gender Classification," *Proceedings of the 1st Conference on Fairness, Accountability and Transparency, PMLR* 81 (2018): 77–91, https://proceedings.mlr.press/v81/buolamwini18a/buolamwini18a.pdf.
3. Buolamwini and Gebru, "Gender Shades."
4. Luis Morales-Navarro, Yasmin B. Kafai, Lauren Vogelstein, Evelyn Yu, and Danaé Metaxa, "Learning About Algorithm Auditing in Five Steps: Scaffolding How High School Youth Can Systematically and Critically Evaluate Machine Learning Applications," *Proceedings of the AAAI Conference on Artificial Intelligence* 39, no. 28 (2025): 29186–29194, https://doi.org/10.1609/aaai.v39i28.35192.
5. Kym L. Worthy, Wayne County Prosecutor, "WCPO Statement in Response to New York Times Article Wrongfully Accused by an Algorithm," press release, June 24, 2020, https://int.nyt.com/data/documenthelper/7046-facial-recognition-arrest/5a6d6d0047295fad363b/optimized/full.pdf.
6. Worthy, "WCPO Statement."
7. Worthy, "WCPO Statement."
8. Kyra Yee, Uthaipon Tantipongpipat, and Shubhanshu Mishra, "Image Cropping on Twitter: Fairness Metrics, Their Limitations, and the Importance of Representation, Design, and Agency," *Proceedings of the ACM on Human–*

Computer Interaction 5, no. CSCW2 (2021): 1–24, https://doi.org/10.1145/3479594.

9. J. Nathan Matias, Austin Hounsel, and Nick Feamster, "Software-Supported Audits of Decision-Making Systems: Testing Google and Facebook's Political Advertising Policies," *Proceedings of the ACM on Human–Computer Interaction* 6, no. CSCW1 (2022): 1–19, https://doi.org/10.1145/3512965.

10. John Ebbert, "BMO'S Salmon: Facebook Starting to See Online Video / TV Ad Budget," *AdExchanger*, January 2, 2013, https://www.adexchanger.com/analysts/bmos-salmon-facebook-starting-to-see-online-video-tv-ad-budget; Jim Edwards, "Here's a Diagram of How Facebook's FBX Ad Exchange Works," *Business Insider*, January 3, 2013, https://www.businessinsider.com/how-facebooks-fbx-ad-exchange-works-2013-1.

11. Dan Salmon, *Exhibit 4. How Facebook Exchange Works: JetBlue as Example*, January 3, 2013, digital image, BMO Capital Markets, https://www.businessinsider.com/how-facebooks-fbx-ad-exchange-works-2013-1.

12. Latanya Sweeney, "Discrimination in Online Ad Delivery," *Communications of the ACM* 56, no. 5 (2013): 44–54, https://doi.org/10.1145/2447976.2447990.

13. Sweeney, "Discrimination in Online Ad Delivery."

14. Christian Sandvig, Kevin Hamilton, Karrie Karahalios, and Cedric Langbort, "Auditing Algorithms: Research Methods for Detecting Discrimination on Internet Platforms," paper presented at Data and Discrimination: Converting Critical Concerns into Productive Inquiry, a preconference at the 64th Annual Meeting of the International Communication Association, Seattle, May 22, 2014, https://websites.umich.edu/~csandvig/research/Auditing%20Algorithms%20--%20Sandvig%20--%20ICA%202014%20Data%20and%20Discrimination%20Preconference.pdf.

15. "Report: Orbitz Steers Mac Users to Pricier Hotels," *CBS News*, June 25, 2012, https://www.cbsnews.com/news/report-orbitz-steers-mac-users-to-pricier-hotels/.

16. Aniko Hannak, Gary Soeller, David Lazer, Alan Mislove, and Christo Wilson, "Measuring Price Discrimination and Steering on E-commerce Web Sites," *Proceedings of the 2014 Conference on Internet Measurement* (2014): 305–318, https://doi.org/10.1145/2663716.2663744.

17. "Which States Have Consumer Data Privacy Laws?," *Bloomberg Law*, accessed March 18, 2024, https://pro.bloomberglaw.com/insights/privacy/state-privacy-legislation-tracker/#row-666b72bb4d398.

18. Lara Groves, Jacob Metcalf, Alayna Kennedy, Briana Vecchione, and Andrew Strait, "Auditing Work: Exploring the New York City Algorithmic Bias

Audit Regime," *Proceedings of the 2024 ACM Conference on Fairness, Accountability, and Transparency* (2024): 1107–1120, https://doi.org/10.1145/3630106.3658959.

19. Jeffrey Dastin, "Insight—Amazon Scraps Secret AI Recruiting Tool that Showed Bias Against Women," *Reuters*, October 10, 2018, https://www.reuters.com/article/idUSKCN1MK0AG/; Nick Statt, "Twitter Tries to Fix Problematic Image Crops by Not Cropping Pictures Anymore," *The Verge*, March 10, 2021, https://www.theverge.com/2021/3/10/22323298/twitter-image-cropping-preview-problematic-racial-bias-testing-fix.

20. Hannah Denham, "IBM's Decision to Abandon Facial Recognition Technology Fueled by Years of Debate," *Washington Post*, June 11, 2020, https://www.washingtonpost.com/technology/2020/06/11/ibm-facial-recognition/.

21. Lena Armstrong, Abbey Liu, Stephen MacNeil, and Danaé Metaxa, "The Silicon Ceiling: Auditing GPT's Race and Gender Biases in Hiring," *Proceedings of the 4th ACM Conference on Equity and Access in Algorithms, Mechanisms, and Optimization* (2024): 1–18, https://doi.org/10.1145/3689904.3694699. For an example of a comparative audit method of several LLMs, see Rina Palta, Julia Angwin, and Alondra Nelson, "How We Tested Leading AI Models Performance on Election Queries," *AI Democracy Projects* and *Proof News*, February 27, 2024, https://www.proofnews.org/how-we-tested-leading-ai-models-performance-on-election-queries/.

22. Armstrong et al., "The Silicon Ceiling."

23. Victor Le Pochat, Laura Edelson, Tom Van Goethem, Wouter Joosen, Damon McCoy, and Tobias Lauinger, "An Audit of Facebook's Political Ad Policy Enforcement," *31st USENIX Security Symposium* (2022): 607–624, https://www.usenix.org/conference/usenixsecurity22/presentation/lepochat.

24. Hong Shen, Alicia DeVos, Motahhare Eslami, and Kenneth Holstein, "Everyday Algorithm Auditing: Understanding the Power of Everyday Users in Surfacing Harmful Algorithmic Behaviors," *Proceedings of the ACM on Human–Computer Interaction* 5, no. CSCW2 (2021): 1–29, https://doi.org/10.1145/3479577.

25. Michelle S. Lam, Mitchell L. Gordon, Danaé Metaxa, Jeffery T. Hancock, James A. Landay, and Michael S. Bernstein, "End-User Audits: A System Empowering Communities to Lead Large-Scale Investigations of Harmful Algorithmic Behavior," *Proceedings of the ACM on Human–Computer Interaction* 6, no. CSCW2 (2022): 1–34, https://doi.org/10.1145/3555625.

26. Julia Angwin and Terry Parris Jr., "Facebook Lets Advertisers Exclude Users by Race," *ProPublica*, October 28, 2016, https://www.propublica.org/article/facebook-lets-advertisers-exclude-users-by-race.

27. Ariel J. Feldman, J. Alex Halderman, and Edward W. Felten, "Security Analysis of the Diebold AccuVote-TS Voting Machine," in *Proceedings of the USENIX Workshop on Accurate Electronic Voting Technology* (USENIX Association, 2007), https://dl.acm.org/doi/10.5555/1323111.1323113.
28. J. Nathan Matias, "Humans and Algorithms Work Together—So Study Them Together," *Nature* 617, no. 7960 (2023): 248–251, https://doi.org/10.1038/d41586-023-01521-z.
29. Kyrstin Wallach, "A Content Analysis of Twitter Use: Factors That Might Increase Music Sales During an Award Show," *Elon Journal of Undergraduate Research in Communications* 5, no. 1 (2014): 1–2, http://www.inquiriesjournal.com/a?id=974.
30. "AI Reviews and Inventory," City of San José, accessed June 6, 2024, https://www.sanjoseca.gov/your-government/departments-offices/information-technology/digital-privacy/ai-reviews-algorithm-register.

Chapter 4

1. Daniel Wiessner, "EEOC Says Workday Must Face Claims that AI Software Is Biased," *Reuters*, April 11, 2024, https://www.reuters.com/legal/transactional/eeoc-says-workday-covered-by-anti-bias-laws-ai-discrimination-case-2024-04-11/.
2. Tanmay Manohar and Andrew Avrin, "How Paramount Leverages Behavioral Assessments for Blockbuster DEI Results," *Harver*, October 4, 2023, https://harver.com/resources/webinars/how-paramount-leverages-behavioral-assessments-for-blockbuster-dei-results/.
3. Lucas Wright, Roxana Mika Muenster, Briana Vecchione, Tianyao Qu, Pika (Senhuang) Cai, Alan Smith, et al., "Null Compliance: NYC Local Law 144 and the Challenges of Algorithm Accountability," *Proceedings of the 2024 ACM Conference on Fairness, Accountability, and Transparency* (2024): 1701–1713, https://doi.org/10.1145/3630106.3658998.
4. Samuel Cortinhas, "Muffin vs Chihuahua," *Kaggle*, last modified December 20, 2022, https://www.kaggle.com/datasets/samuelcortinhas/muffin-vs-chihuahua-image-classification.
5. Richard Berk, Hoda Heidari, Shahin Jabbari, Michael Kearns, and Aaron Roth, "Fairness in Criminal Justice Risk Assessments: The State of the Art," *Sociological Methods & Research* 50, no. 1 (2018): 3–44, https://doi.org/10.1177/0049124118782533.
6. What "accuracy" means in the evaluation of AI is a nuanced question. Some definitions cover up AI's deficiencies. See: Gabriel Grill, "Constructing Capabilities: The Politics of Testing Infrastructures for Generative AI," *Proceedings of*

the 2024 ACM Conference on Fairness, Accountability, and Transparency (2024): 1838–1849, https://doi.org/10.1145/3630106.3659009.
7. "Select Issues: Assessing Adverse Impact in Software, Algorithms, and Artificial Intelligence Used in Employment Selection Procedures Under Title VII of the Civil Rights Act of 1964," US Equal Employment Opportunity Commission, May 18, 2023, https://web.archive.org/web/20241127052514/https://www.eeoc.gov/laws/guidance/select-issues-assessing-adverse-impact-software-algorithms-and-artificial.
8. Uniform Guidelines on Employee Selection Procedures (1978), 29 C.F.R § 1607.4.
9. *Summary of Bias Audit Results: Audit of the HackerRank's Image Analysis System for New York City's Local Law 144* (BABL AI, July 22, 2024), https://support.hackerrank.com/hc/en-us/articles/18059959675539-Summary-of-Bias-Audit-Results-HackerRank-s-Image-Analysis-System.
10. Kimmo Karkkainen and Jungseock Joo, "FairFace: Face Attribute Dataset for Balanced Race, Gender, and Age for Bias Measurement and Mitigation," *Proceedings of the IEEE/CVF Winter Conference on Applications of Computer Vision* (2021): 1548–1558, https://openaccess.thecvf.com/content/WACV2021/html/Karkkainen_FairFace_Face_Attribute_Dataset_for_Balanced_Race_Gender_and_Age_WACV_2021_paper.html.
11. Jeffrey Dastin, "Insight—Amazon Scraps Secret AI Recruiting Tool that Showed Bias Against Women," *Reuters*, October 10, 2018, https://www.reuters.com/article/idUSKCN1MK0AG/.
12. Lara Groves, Jacob Metcalf, Alayna Kennedy, Briana Vecchione, and Andrew Strait, "Auditing Work: Exploring the New York City Algorithmic Bias Audit Regime," *Proceedings of the 2024 ACM Conference on Fairness, Accountability, and Transparency* (2024): 1107–1120, https://doi.org/10.1145/3630106.3658959.
13. ACLU and NYCLU, "Tracking Automated Employment Decision Tool Bias Audits," GitHub repository, created December 10, 2023, https://github.com/aclu-national/tracking-ll144-bias-audits?tab=readme-ov-file.
14. *Summary of Bias Audit Results: Audit of the pymetrics Soft Skills Platform for New York City's Local Law 144* (BABL AI, June 29, 2023), https://www.paramount.com/sites/g/files/dxjhpe226/files/2023-07/Harver_Pymetrics-Final_Audit_Summary-2023-06-29.pdf.
15. *AEDT Bias Audit—SmartAssistant* (ConductorAI, July 11, 2023), https://web.archive.org/web/20240922222751/https://www.conductorai.co/nyc-144-audits/smartassistant.

Chapter 5

1. For example, in 47 U.S.C. §338(d), a law about television "channel positioning," a carrier "must provide access . . . in a nondiscriminatory manner on any navigational device, on-screen program guide, or menu."

2. Nikhil Sonnad, "Google Translate's Gender Bias Pairs 'He' with 'Hardworking' and 'She' with Lazy, and Other Examples," *Quartz*, last updated July 20, 2022, https://qz.com/1141122/google-translates-gender-bias-pairs-he-with-hardworking-and-she-with-lazy-and-other-examples.

3. Tolga Bolukbasi, Kai-Wei Chang, James Zou, Venkatesh Saligrama, and Adam Kalai, "Man Is to Computer Programmer as Woman Is to Homemaker? Debiasing Word Embeddings," *Proceedings of the 30th International Conference on Neural Information Processing Systems* (2016): 4356–4364, https://proceedings.neurips.cc/paper_files/paper/2016/file/a486cd07e4ac3d270571622f4f316ec5-Paper.pdf.

4. Joy Buolamwini and Timnit Gebru, "Gender Shades: Intersectional Accuracy Disparities in Commercial Gender Classification," *Proceedings of the 1st Conference on Fairness, Accountability and Transparency, PMLR* 81 (2018): 77–91, https://proceedings.mlr.press/v81/buolamwini18a/buolamwini18a.pdf.

5. Inioluwa Deborah Raji and Joy Buolamwini, "Actionable Auditing: Investigating the Impact of Publicly Naming Biased Performance Results of Commercial AI Products," *Proceedings of the 2019 AAAI/ACM Conference on AI, Ethics, and Society* (2019): 429–435, https://doi.org/10.1145/3306618.3314244.

6. Juhi Kulshrestha, Motahhare Eslami, Johnnatan Messias, Muhammad Bilal Zafar, Saptarshi Ghosh, Krishna P. Gummadi, et al., "Quantifying Search Bias: Investigating Sources of Bias for Political Searches in Social Media," *Proceedings of the 2017 ACM Conference on Computer Supported Cooperative Work and Social Computing* (2017): 417–432, https://doi.org/10.1145/2998181.2998321.

7. Tom Simonite, "When It Comes to Gorillas, Google Photos Remains Blind," *Wired*, January 11, 2018, https://www.wired.com/story/when-it-comes-to-gorillas-google-photos-remains-blind/; James Vincent, "Google 'Fixed' Its Racist Algorithm by Removing Gorillas from Its Image-Labeling Tech," *The Verge*, January 12, 2018, https://www.theverge.com/2018/1/12/16882408/google-racist-gorillas-photo-recognition-algorithm-ai.

8. Nico Grant and Kashmir Hill, "Google's Photo App Still Can't Find Gorillas. And Neither Can Apple's," *New York Times*, May 22, 2023, https://www.nytimes.com/2023/05/22/technology/ai-photo-labels-google-apple.html.

9. Grant and Hill, "Google's Photo App Still Can't Find Gorillas."

10. Julia Angwin and Terry Parris Jr., "Facebook Lets Advertisers Exclude Users by Race," *ProPublica*, October 28, 2016, https://www.propublica.org/article/facebook-lets-advertisers-exclude-users-by-race.

11. "Improving Enforcement and Promoting Diversity: Updates to Ads Policies and Tools," Facebook, February 8, 2017, https://about.fb.com/news/2017/02/improving-enforcement-and-promoting-diversity-updates-to-ads-policies-and-tools/.

12. Julia Angwin, Ariana Tobin, and Madeleine Varner, "Facebook (Still) Letting Housing Advertisers Exclude Users by Race," *ProPublica*, November 21, 2017, https://www.propublica.org/article/facebook-advertising-discrimination-housing-race-sex-national-origin.

13. Sheryl Sandberg, "Doing More to Protect Against Discrimination in Housing, Employment and Credit Advertising," Meta, March 19, 2019, https://about.fb.com/news/2019/03/protecting-against-discrimination-in-ads/.

14. See, for example, Muhammad Ali, Piotr Sapiezynski, Miranda Bogen, Aleksandra Korolova, Alan Mislove, and Aaron Rieke, "Discrimination Through Optimization: How Facebook's Ad Delivery Can Lead to Biased Outcomes," *Proceedings of the ACM on Human–Computer Interaction* 3, no. CSCW (2019): 1–30, https://doi.org/10.1145/3359301.

15. Hilke Schellmann, "Auditors Are Testing Hiring Algorithms for Bias, but There's No Easy Fix," *MIT Technology Review*, February 11, 2021, https://www.technologyreview.com/2021/02/11/1017955/auditors-testing-ai-hiring-algorithms-bias-big-questions-remain/.

16. Buolamwini and Gebru, "Gender Shades."

17. Raji and Buolamwini, "Actionable Auditing."

18. Nari Johnson, Sanika Moharana, Christina Harrington, Nazanin Andalibi, Hoda Heidari, and Motahhare Eslami, "The Fall of an Algorithm: Characterizing the Dynamics Toward Abandonment," *Proceedings of the 2024 ACM Conference on Fairness, Accountability, and Transparency* (2024): 337–358, https://doi.org/10.1145/3630106.3658910.

19. Mark Puente, "LAPD Ends Another Data-Driven Crime Program Touted to Target Violent Offenders," *Los Angeles Times*, April 12, 2019, https://www.latimes.com/local/lanow/la-me-laser-lapd-crime-data-program-20190412-story.html; Kathleen Foody, "Chicago Police End Effort to Predict Gun Offenders, Victims," *AP*, January 23, 2020, https://apnews.com/general-news-41f75b783d796b80815609e737211cc6.

20. Adam Harvey, "HyperFace," developed for Hyphen-Labs, 2017, https://adam.harvey.studio/hyperface/.

21. Kevin Eykholt, Ivan Evtimov, Earlence Fernandes, Bo Li, Amir Rahmati, Florian Tramèr, et al., "Physical Adversarial Examples for Object Detectors," in *Proceedings of the 12th USENIX Conference on Offensive Technologies* (USENIX Association, 2018), https://doi.org/10.48550/arXiv.1807.07769.

22. Lauren Neergaard, "A Racially Biased Test Kept Thousands of Black People from Getting a Kidney Transplant," *PBS*, April 1, 2024, https://www.pbs.org/newshour/nation/a-racially-biased-test-kept-thousands-of-black-people-from-getting-a-kidney-transplant.

23. Sachin Waikar, "Evolution of an Algorithm: Lessons from the Kidney Allocation System," *Stanford University Human-Centered Artificial Intelligence*, April 27, 2020, https://hai.stanford.edu/news/evolution-algorithm-lessons-kidney-allocation-system.

24. "Reddit Rules," Reddit, accessed June 6, 2024, https://www.redditinc.com/policies/reddit-rules.

25. "Community Standards," Meta, accessed June 6, 2024, https://transparency.meta.com/policies/community-standards/.

26. Danaé Metaxa, Joon Sung Park, James A. Landay, and Jeff Hancock, "Search Media and Elections: A Longitudinal Investigation of Political Search Results," *Proceedings of the ACM on Human–Computer Interaction* 3, no. CSCW (2019): 1–17, https://doi.org/10.1145/3359231.

27. Kate Evans and Robert Koulish, "Manipulating Risk: Immigration Detention Through Automation," *Lewis & Clark Law Review* 24, no. 3 (2020): 789–855, https://scholarship.law.duke.edu/cgi/viewcontent.cgi?article=6692&context=faculty_scholarship.

28. *Big Data: A Report on Algorithmic Systems, Opportunity, and Civil Rights* (Executive Office of the President, 2016), https://obamawhitehouse.archives.gov/sites/default/files/microsites/ostp/2016_0504_data_discrimination.pdf.

29. *Blueprint for an AI Bill of Rights: Making Automated Systems Work for the American People* (White House Office of Science and Technology Policy, 2022), https://web.archive.org/web/20241127052631/https://www.whitehouse.gov/wp-content/uploads/2022/10/Blueprint-for-an-AI-Bill-of-Rights.pdf; Exec. Order No. 14110, 88 Fed. Reg. 75191 (October 30, 2023).

30. Algorithmic Accountability Act of 2022, H.R. 6580, 117th Cong. (2022).

31. NYC Consumer and Worker Protection, "Automated Employment Decision Tools: Frequently Asked Questions," NYC Department of Consumer and Worker Protection, June 29, 2023, https://www.nyc.gov/assets/dca/downloads/pdf/about/DCWP-AEDT-FAQ.pdf.

32. Liapes v. Facebook, Inc., 95 Cal.App.5th 910 (Cal. Ct. App. 2023); Ali et al., "Discrimination Through Optimization."

33. Havens Realty Corp. v. Coleman, 455 U.S. 363 (1982).

34. US Department of Justice, "Justice Department Secures Groundbreaking Settlement Agreement with Meta Platforms, Formerly Known as Facebook, to Resolve Allegations of Discriminatory Advertising," press release 22-650, June 21, 2022, updated February 6, 2025, https://www.justice.gov/opa/pr/justice-department-secures-groundbreaking-settlement-agreement-meta-platforms-formerly-known.

35. United States v. Meta Platforms, Inc., 22-cv-05187 (S.D. N.Y. 2022).

36. "K.W. v. Armstrong," ACLU Idaho, January 18, 2012, https://www.acluidaho.org/en/cases/kw-v-armstrong.

37. Jay Stanley, "Pitfalls of Artificial Intelligence Decisionmaking Highlighted in Idaho ACLU Case," ACLU, June 2, 2017, https://www.aclu.org/news/privacy-technology/pitfalls-artificial-intelligence-decisionmaking-highlighted-idaho-aclu-case.

38. *Report on Apple Card Investigation* (New York State Department of Financial Services, 2021), https://www.dfs.ny.gov/system/files/documents/2021/03/rpt_202103_apple_card_investigation.pdf.

39. Ian Carlos Campbell, "The Apple Card Doesn't Actually Discriminate Against Women, Investigators Say," *The Verge*, March 23, 2021, https://www.theverge.com/2021/3/23/22347127/goldman-sachs-apple-card-no-gender-discrimination.

40. Molly Weisner, "Data Breaches, Led by USPS, OPM, Cost Governments $26 Billion," *Federal Times*, December 30, 2022, https://www.federaltimes.com/it-networks/2022/12/30/data-breaches-led-by-usps-opm-cost-governments-26-billion/.

41. Norman Isaac Silber, *Test and Protest: The Influence of Consumers Union* (Holmes and Meier, 1983).

42. Consumer Reports, "Consumer Reports Study Finds Significant Obstacles to Exercising California Privacy Rights," press release, October 1, 2020, https://advocacy.consumerreports.org/press_release/consumer-reports-study-finds-significant-obstacles-to-exercising-california-privacy-rights/.

43. Consumer Reports, "Consumer Reports Introduces Free 'Permission Slip by CR' App to Empower Consumers to Take Back Control of Their Personal Data," press release, October 3, 2023, https://www.consumerreports.org/media-room/press-releases/2023/10/consumer-reports-introduces-free-permission-slip-by-cr-app-to-empower-consumers-to-take-back-control-of-their-personal-data/.

44. Raji and Buolamwini, "Actionable Auditing."

45. Frances Cherry and Marc Bendick Jr., "Making It Count: Discrimination Auditing and the Activist Scholar Tradition," in *Audit Studies: Behind the Scenes with Theory, Method, and Nuance*, ed. S. Michael Gaddis (Springer, 2018), 45–62.

46. Kate Wells, "An Eating Disorders Chatbot Offered Dieting Advice, Raising Fears About AI in Health," *NPR*, June 9, 2023, https://www.npr.org/sections/health-shots/2023/06/08/1180838096/an-eating-disorders-chatbot-offered-dieting-advice-raising-fears-about-ai-in-hea; "NEDA Suspends AI Chatbot for Giving Harmful Eating Disorder Advice," *Psychiatrist.com*, June 5, 2023, https://www.psychiatrist.com/news/neda-suspends-ai-chatbot-for-giving-harmful-eating-disorder-advice/; Ragen Chastain, "Eating Disorders Support Chat-Bot Promotes Weight Loss," *Weight and Healthcare* (blog), May 31, 2023, https://weightandhealthcare.substack.com/p/nedas-union-busting-eating-disorders.

47. Susan Morse, "Study Finds Racial Bias in Optum Algorithm," *Healthcare Finance*, October 25, 2019, https://www.healthcarefinancenews.com/news/study-finds-racial-bias-optum-algorithm.

48. Melanie Evans and Anna Wilde Mathews, "New York Regulator Probes UnitedHealth Algorithm for Racial Bias," *Wall Street Journal*, October 26, 2019, https://www.wsj.com/articles/new-york-regulator-probes-unitedhealth-algorithm-for-racial-bias-11572087601.

Chapter 6

1. Stephanie Marken and Tara Nicola, "Three in Four Americans Believe AI Will Reduce Jobs," *Gallup* (blog), September 13, 2023, https://news.gallup.com/opinion/gallup/510635/three-four-americans-believe-reduce-jobs.aspx; "There Is Bipartisan Concern About the Use of AI in the 2024 Elections," AP-NORC Center for Public Affairs Research, November 3, 2023, https://apnorc.org/projects/there-is-bipartisan-concern-about-the-use-of-ai-in-the-2024-elections; Michelle Faverio and Alec Tyson, "What the Data Says About Americans' Views of Artificial Intelligence," Pew Research Center, November 21, 2023, https://www.pewresearch.org/short-reads/2023/11/21/what-the-data-says-about-americans-views-of-artificial-intelligence/.

2. Jonathan Taplin, *Move Fast and Break Things: How Facebook, Google, and Amazon Cornered Culture and Undermined Democracy* (Little, Brown, 2017).

3. Ben Barry, "Facebook Posters 2010–2013," Office of Ben Barry, accessed June 6, 2024, https://v1.benbarry.com/project/facebook-posters.

4. Henry Blodget, "Facebook Strategy Revealed: Move Fast and Break Things!," *Business Insider*, March 1, 2010, https://www.businessinsider.com/henry-blodget-innovation-highlights-2010-2.

5. Isobel Asher Hamilton, "Mark Zuckerberg's New Values for Meta Show He Still Hasn't Truly Let Go of 'Move Fast and Break Things,'" *Business Insider*, February 16, 2022, https://www.businessinsider.com/meta-mark-zuckerberg-new-values-move-fast-and-break-things-2022-2.

6. Charles Perrow, *Normal Accidents: Living with High-Risk Technologies*, updated ed. (Basic Books, 1984; Princeton University Press, 1999). Citations refer to the Princeton edition.

7. *After the Dead Are Counted: U.S. and Pakistani Responsibilities to Victims of Drone Strikes* (Open Society Foundations, 2014), https://www.opensocietyfoundations.org/publications/after-dead-are-counted-us-and-pakistani-responsibilities-victims-drone-strikes.

8. Christian Grothoff and J. M. Porup, "The NSA's SKYNET Program May Be Killing Thousands of Innocent People," *Ars Technica*, February 16, 2016, https://arstechnica.com/information-technology/2016/02/the-nsas-skynet-program-may-be-killing-thousands-of-innocent-people/.

9. Sergio García Magariño, "What Is Hamas? Seven Key Questions Answered," *The Conversation*, October 11, 2023, https://theconversation.com/what-is-hamas-seven-key-questions-answered-215391; Yuval Abraham, "'Lavender': The AI Machine Directing Israel's Bombing Spree in Gaza," *+972 Magazine*, April 3, 2024, https://www.972mag.com/lavender-ai-israeli-army-gaza/.

10. "Knuth Reward Check," Wikipedia, last modified December 25, 2023, https://en.wikipedia.org/wiki/Knuth_reward_check.

11. Kurt W. Beyer, *Grace Hopper and the Invention of the Information Age* (MIT Press, 2009).

12. *Identifying Security Vulnerabilities in Department of Defense Websites—Hack the Pentagon* (United States Digital Service, 2016), https://www.usds.gov/report-to-congress/2016/hack-the-pentagon/.

13. Hayden Field, "Behind the Scenes: How Twitter Decided to Open Up Its Image-Cropping Algorithm to the Public," *Tech Brew*, September 27, 2021, https://www.emergingtechbrew.com/stories/2021/09/27/behind-the-scenes-of-twitter-s-decision-to-open-up-its-image-cropping-algorithm-to-researchers.

14. Bogdan Kulynych, "How to Become More Salient? Surfacing Representation Biases of the Saliency Prediction Model," Github repository, created August 11, 2021, https://github.com/bogdan-kulynych/saliency_bias.

15. "Twitter Drops Image-Cropping Algorithm After Finding Bias," *TechXplore*, May 19, 2021, https://techxplore.com/news/2021-05-twitter-image-cropping-algorithm-bias.html#google_vignette.

16. Corinne Cath, Mark Latonero, Vidushi Marda, and Roya Pakzad, "Leap of FATE: Human Rights as a Complementary Framework for AI Policy and

Practice," *Proceedings of the 2020 Conference on Fairness, Accountability, and Transparency* (2020): 702, https://doi.org/10.1145/3351095.3375665.

17. Dunstan Allison-Hope, "Our Human Rights Impact Assessment of Facebook in Myanmar," *BSR* (blog), November 5, 2018, https://www.bsr.org/en/blog/facebook-in-myanmar-human-rights-impact-assessment.

18. Alex Warofka, "An Independent Assessment of the Human Rights Impact of Facebook in Myanmar," Meta, November 5, 2018, https://about.fb.com/news/2018/11/myanmar-hria/.

19. Esha Bhandari, "Uncovering Online Discrimination When Faced with Legal Uncertainty and Corporate Power," in *Feminist Cyberlaw*, ed. Meg Leta Jones and Amanda Levendowski (University of California Press, 2024), 117–128, https://doi.org/10.2307/jj.14086449.12.

20. "Sandvig v. Barr—Challenge to CFAA Prohibition on Uncovering Racial Discrimination Online," ACLU, last modified May 22, 2019, https://www.aclu.org/cases/sandvig-v-barr-challenge-cfaa-prohibition-uncovering-racial-discrimination-online.

21. Bhandari, "Uncovering Online Discrimination."

22. *Van Buren v. United States* cited the brief for Kyratso Karahalios et al. as amici curiae; Van Buren v. United States, 141 S. Ct. 1648 (2021) at 7.

23. Komal S. Patel, "Testing the Limits of the First Amendment: How Online Civil Rights Testing Is Protected Speech Activity," *Columbia Law Review* 118, no. 5 (2018): 1473–1516, https://columbialawreview.org/content/testing-the-limits-of-the-first-amendment-how-online-civil-rights-testing-is-protected-speech-activity/.

24. Hannah L. Semigran, Jeffrey A. Linder, Courtney Gidengil, and Ateev Mehrotra, "Evaluation of Symptom Checkers for Self Diagnosis and Triage: Audit Study," *BMJ* 351 (2015), https://doi.org/10.1136/bmj.h3480.

25. David C. Vladeck, "Machines Without Principals: Liability Rules and Artificial Intelligence," *Washington Law Review* 89, no. 1 (2014): 117–150, https://digitalcommons.law.uw.edu/wlr/vol89/iss1/6.

26. Nilsson v. General Motors LLC (N.D. Cal. Jan. 22, 2018).

27. Rena Li, Sara Kingsley, Chelsea Fan, Proteeti Sinha, Nora Wai, Jaimie Lee, et al., "Participation and Division of Labor in User-Driven Algorithm Audits: How Do Everyday Users Work Together to Surface Algorithmic Harms?," *Proceedings of the 2023 CHI Conference on Human Factors in Computing Systems* (2023): 1–19, https://doi.org/10.1145/3544548.3582074.

28. "Helping Researchers Understand Online Behavior," National Internet Observatory, accessed June 6, 2024, https://nationalinternetobservatory.org/.

29. Johana Bhuiyan, "WhatsApp's AI Shows Gun-Wielding Children When Prompted with 'Palestine,'" *The Guardian*, November 3, 2023, https://www.theguardian.com/technology/2023/nov/02/whatsapps-ai-palestine-kids-gun-gaza-bias-israel.

30. Amos Toh, *Automated Neglect: How the World Bank's Push to Allocate Cash Assistance Using Algorithms Threatens Rights* (Human Rights Watch, 2023), https://www.hrw.org/report/2023/06/13/automated-neglect/how-world-banks-push-allocate-cash-assistance-using-algorithms.

31. Eileen Guo, Gabriel Geiger, and Justin-Casimir Braun, "Inside Amsterdam's High-Stakes Experiment to Create Fair Welfare AI," *MIT Technology Review*, June 11, 2025, https://www.technologyreview.com/2025/06/11/1118233/amsterdam-fair-welfare-ai-discriminatory-algorithms-failure/.

32. Prabhakar Raghavan, "Gemini Image Generation Got It Wrong. We'll Do Better," *Google* (blog), February 23, 2024, https://blog.google/products/gemini/gemini-image-generation-issue/.

33. Rebecca MacKinnon, *Consent of the Networked: The Worldwide Struggle for Internet Freedom* (Basic Books, 2012); Ellery Roberts Biddle, "The New Gatekeepers: Will Google Decide How We Remember Syria's Civil War?," Monument Lab and *Global Voices*, February 19, 2019, https://monumentlab.com/bulletin/the-new-gatekeepers-syrian-war.

FURTHER READING

Amoore, Louise. *Cloud Ethics: Algorithms and the Attributes of Ourselves and Others*. Duke University Press, 2020.

Benjamin, Ruha. *Race After Technology: Abolitionist Tools for the New Jim Code*. Polity Press, 2019.

Blank, Rebecca M., Marilyn Dabady, and Constance F. Citro, eds. *Measuring Racial Discrimination*. National Academies Press, 2004. See esp. chap. 6, "Experimental Methods for Assessing Discrimination."

Buolamwini, Joy. *Unmasking AI: My Mission to Protect What Is Human in a World of Machines*. Random House, 2023.

Christian, Brian. *The Alignment Problem: Machine Learning and Human Values*. W. W. Norton, 2020.

D'Ignazio, Catherine, and Lauren F. Klein. *Data Feminism*. MIT Press, 2020.

Eubanks, Virginia. *Automating Inequality: How High-Tech Tools Profile, Police, and Punish the Poor*. St. Martin's Press, 2018.

Gaddis, S. Michael, ed. *Audit Studies: Behind the Scenes with Theory, Method, and Nuance*. Springer, 2018.

International Panel on the Information Environment. *Recommendations for a Global AI Auditing Framework: Summary of Standards and Features*. SFP2024.2. IPIE, 2024. https://www.ipie.info/research/sfp2024-2.

Levy, Karen. *Data Driven: Truckers, Technology, and the New Workplace Surveillance*. Princeton University Press, 2023.

Metaxa, Danaé, Joon Sung Park, Ronald E. Robertson, Karrie Karahalios, Christo Wilson, Jeff Hancock, and Christian Sandvig. *Auditing Algorithms: Understanding Algorithmic Systems from the Outside In*. Now Publishers, 2021.

Noble, Safiya Umoja. *Algorithms of Oppression: How Search Engines Reinforce Racism*. NYU Press, 2018.

O'Neil, Cathy. *Weapons of Math Destruction: How Big Data Increases Inequality and Threatens Democracy*. Crown, 2016.

Pasquale, Frank. *The Black Box Society: The Secret Algorithms That Control Money and Information*. Harvard University Press, 2015.

Seaver, Nick. *Computing Taste: Algorithms and the Makers of Music Recommendation*. University of Chicago Press, 2022.

INDEX

The **MARQUAND HOUSE COLLECTIVE** comprises eleven experts in AI auditing spanning computing, law, policy, social science, and journalism. Members coined the term "algorithm audit" in a 2014 publication. The full group convened in 2024 at Marquand House in Princeton, New Jersey.

MARC AIDINOFF is an assistant professor of history of science at Harvard University. His forthcoming book, *Rebooting Liberalism: The Computerization of the Social Contract from 1974 to 2004*, examines the computerization of government services and the changing expectations of citizenship. He has previously served as a science and technology policy advisor in the Biden-Harris and Obama-Biden administrations.

LENA ARMSTRONG is a computer science PhD student at Harvard University. Her research focuses on human–computer interaction, algorithmic justice, and AI auditing and has been awarded a National Science Foundation Graduate Research Fellowship. She leverages interdisciplinary approaches to investigate bias and opacity in automated systems, create frameworks to audit AI, and discover how to empower people through more transparent and equitable systems.

ESHA BHANDARI is director of the ACLU Speech, Privacy, and Technology Project, where she works on litigation and advocacy to protect freedom of expression and privacy rights in the digital age. She litigated the case *Sandvig v. Barr*, a First Amendment challenge to the Computer Fraud and Abuse Act on behalf of online discrimination researchers. Bhandari is an adjunct professor of clinical law at New York University School of Law, where she coteaches the Technology, Law, and Policy Clinic. She contributed a chapter to the treatise *Feminist Cyberlaw* on the legal landscape for digital journalism and research.

ELLERY ROBERTS BIDDLE is a Philadelphia-based journalist who covers technology and the ways it affects people's lives and rights around the world. Her work has earned accolades from the Society of Professional Journalists, the Fetisov Journalism Awards, the European Journalism Centre, and the Orwell Foundation. A former fellow at Harvard's Berkman Klein Center for Internet and Society, and former managing editor at *Coda Story*, her writing has appeared in *The New York Times*, *The Guardian*, *Slate*, and *Compiler News*, among other outlets.

MOTAHHARE ESLAMI is an assistant professor at the School of Computer Science, Human-Computer Interaction Institute (HCII), and Software and Societal Systems Department (S3D) at Carnegie Mellon University. Her research is broadly in human–computer interaction, AI, social computing, and data mining, and aims to engage diverse stakeholders in the design and governance of AI systems, ensuring that those most affected by AI systems are given a voice in shaping these technologies. Eslami was named one of the 100 Brilliant Women in AI Ethics, and her work has been recognized with several awards at top-tier ACM and AAAI conferences.

KARRIE KARAHALIOS is a professor of media arts and sciences at the Massachusetts Institute of Technology, where she directs the Social Algorithms Lab. Karahalios's award-winning research focuses on AI and community-centered computing. She studies and builds sociotechnical systems, exploring algorithm awareness, algorithmically mediated social media feeds, content moderation systems, accessibility, group dynamics, and just infrastructures, to name a few. She coauthored the original paper on algorithm auditing and was a plaintiff in *Sandvig v. Barr*, whose outcome acknowledged auditing for civil rights should not be criminal. She was previously at the University of Illinois where she founded the Center for Just Infrastructures.

J. NATHAN MATIAS is a computer scientist and social scientist who organizes citizen behavioral science for a safer, fairer, more understanding internet. A Guatemalan American, Matias is founder of the Citizens and Technology Lab at Cornell University. His award-winning scientific research, published in *Nature*, *Science*, *PNAS*, and other leading journals, has enlisted the public to audit social technologies and measurably improve the digital lives of hundreds of millions of people worldwide. Matias also cofounded the Coalition for Independent Technology Research, which works to support and defend independent research on technology and society.

DANAÉ METAXA is Raj and Neera Singh Assistant Professor of Computer Science at the University of Pennsylvania, with a secondary appointment in the Annenberg School for Communication. They have coauthored one other book, *Auditing Algorithms: Understanding Algorithmic Systems from the Outside In*, the authoritative academic review on AI auditing. Metaxa researches algorithmic bias, representation, and justice in sociotechnical systems, and works with legal, industry, and policy groups including the New York City Commission on Human Rights and the American Civil Liberties Union.

ALONDRA NELSON is the Harold F. Linder Professor in the School of Social Science at the Institute for Advanced Study, where she leads the Science, Technology, and Social Values Lab. Nelson is the author of award-winning books, including *The Social Life of DNA*. She led development of the White House "Blueprint for an AI Bill of Rights," served on the United Nations High-level Advisory Body on AI, and was recognized on the inaugural *TIME* 100 list of the most influential people in AI.

CHRISTIAN SANDVIG is the director of the Center for Ethics, Society, and Computing (ESC) and H. Marshall McLuhan Collegiate Professor of Information at the University of Michigan. He coauthored the original academic paper on algorithm auditing, coining the term, and was a plaintiff in *Sandvig v. Barr*, a lawsuit that changed the definition of hacking in the United States. He previously worked as a computer programmer at a Fortune 500 company, for the government, and at a tech start-up, and was a faculty member at the University of Illinois and Oxford University. His research focuses on the consequences of computer systems that automatically curate and organize culture and information.

KRISTEN VACCARO is an assistant professor of computer science & engineering at the University of California San Diego, where she is also a member of the Design Lab. Her research explores how to design machine learning systems—particularly those found in social media, like content moderation and newsfeeds—to be understandable, trustworthy, and just. This research has been published in top conferences in human–computer interaction, including CHI, CSCW, and UIST. She earned her PhD in computer science from the University of Illinois Urbana-Champaign and has also worked as a researcher at the MITRE Corporation.

Publisher contact:
The MIT Press
Massachusetts Institute of Technology
77 Massachusetts Avenue, Cambridge, MA 02139
mitpress.mit.edu

EU Authorised Representative:
Easy Access System Europe, Mustamäe tee 50,
10621 Tallinn, Estonia
gpsr.requests@easproject.com

Printed by Integrated Books International,
United States of America